Simple Roots

The Minimalist
Guide to Modern
Homesteading

NICCI BROCHARD
&
DR. BEN CHUBA

Book Formatting by: *Kamala Pen*

Book Cover Design by: *Billy Design*

CROSSBORDER

New York, London, Quebec

CONTENTS

FORWARD

What a world consumed by clutter. We mean both physical and mental. *Simple Roots* is a call to return to what truly matters. Modern life often pulls us in countless directions, leaving us overwhelmed, overextended, and disconnected from the land beneath our feet. But there's another way. Through minimalist principles and intentional living, modern homesteading offers a path to reclaiming your time, space, and well-being. This book is about creating a life that is both sustainable and fulfilling, free from the noise of modern consumerism.

For whatever reason you are drawn to homesteading, for self-sufficiency, sustainability, or simply a slower pace of life, *Simple Roots* will guide you in transforming your home, your habits, and your mindset. It's time to strip away the excess and build a life grounded in simplicity, purpose, and connection to the earth. The roots of a meaningful life are simpler than you think.

Ben and I (Nicci) want to that you in advance for picking up this book to read. We know you will find what you are looking for.

Introduction
The Minimalist's Guide to Modern Homesteading

In many ways, we are striving for more, more possessions, more responsibilities, more distractions. It's easy to feel overwhelmed by the noise of modern life. Amidst the endless chase for material success, many are beginning to realize that true happiness and fulfillment often lie in the simplicity of what we already have. This book, *Simple Roots: The Minimalist's Guide to Modern Homesteading*, is designed for those who wish to return to the basics, finding peace, purpose, and sustainability in the simplicity of homesteading while embracing the principles of minimalism.

Homesteading, often seen as an outdated or rustic way of life, is rapidly being redefined in today's fast-paced world. Modern homesteading no longer means abandoning city life or living completely off the grid, but it does embrace the idea of taking responsibility for one's own well-being. It's about creating a self-sustaining lifestyle that values quality over quantity, and focusing on what really matters, family, community, and a closer connection to nature. Whether you're looking to grow your own food, learn how to preserve it, or simply create a simpler, more fulfilling lifestyle, this book will guide you through every step of the way.

Minimalism has taken the world by storm in recent years, and it's easy to see why. Minimalism teaches us to shed the excess—the things, the habits, and the pressures—that prevent us from living fully and enjoying the beauty of simplicity. By adopting a minimalist mindset, we can focus on what brings us true joy, cultivate mindfulness in everyday tasks, and be more intentional about how we live our lives. In *Simple Roots*, we explore how minimalism aligns perfectly with the principles of homesteading, creating a space where sustainability, simplicity, and mindfulness intersect.

At its core, modern homesteading is about embracing a way of life that is self-sufficient, sustainable, and enriching. It's about cutting out the noise of consumerism and reconnecting with the earth in a way that supports both your well-being and the planet's health. Whether you live on a large plot of land in the countryside or in an urban setting, modern homesteading offers tools and strategies to integrate more sustainable practices into your life. It's about learning how to care for yourself and your family with your own hands, cultivating food, and being more mindful of your consumption. It's about creating a home that nourishes not only your body but also your soul.

In this guide, you will be introduced to simple yet effective ways to implement homesteading practices into your daily life, no matter your current living situation. From growing your own herbs, vegetables, and fruits, to learning how to preserve food through canning or fermenting, you will be empowered to take small, manageable steps toward a more sustainable lifestyle. Alongside these practical skills, you'll also learn how to minimize your environmental impact, reduce waste, and create a harmonious and peaceful home environment.

The principles of minimalism and homesteading are inextricably linked, and together, they create a powerful framework for building a life that aligns with your values. By focusing on what truly matters and simplifying your lifestyle, you open up space for creativity, connection, and freedom. This book will encourage you to slow down, reconnect with the earth, and cultivate a life that is both purposeful and joyful. Whether you're looking to reduce your ecological footprint, save money, or simply live with more intention, *Simple Roots* offers a step-by-step approach to embracing a simpler, more meaningful existence.

Modern homesteading is about making choices that nourish the body, mind, and spirit, and it's a path that anyone can begin to walk—no matter where you start. In the following chapters, you'll find practical advice, inspiration, and tools to help you create a homesteading lifestyle that works for you. As you explore the pages of *Simple Roots*, you'll be reminded that the beauty of life often lies in the simple things—and it's in those small, intentional choices that we can truly cultivate happiness and well-being.

So, let's begin this journey together. Let's root ourselves in the simple joys of life, cultivate a home that supports our values, and embrace a life of intention, sustainability, and peace. Welcome to the minimalist's guide to modern homesteading.

CHAPTER 1

The Call to Simplicity

I n a world filled with endless distractions, consumption, and the pressure to always do more, many are choosing to return to the basics. Minimalism and homesteading are no longer concepts reserved for the isolated rural farmer or the austere lifestyle of a remote individual. Today, more people are combining these concepts in what is known as minimalist homesteading—a lifestyle that emphasizes simplicity, sustainability, and intentional living. This chapter will explore why more people are turning to minimalist homesteading, the distinction between minimalism and deprivation, and how combining modern homesteading with intentional living offers a pathway to a more meaningful life.

Why People Are Choosing Minimalist Homesteading

Minimalist homesteading is a response to the modern world's overconsumption and disconnection from nature. It merges the values of minimalism—living with less but more intentionally—with the age-old practices of homesteading, which revolve around self-sufficiency, sustainability, and a deeper connection to the earth.

1.   Reconnecting with Nature

Modern life, especially in urbanized areas, often disconnects individuals from the natural world. Many people, especially in fast-paced cities, are removed from the processes of food production, energy consumption, and other fundamental aspects of life that sustain human existence. Minimalist homesteading invites people to reconnect with the earth— whether by planting a garden, raising animals, or practicing sustainable farming techniques. This reconnection brings people closer to the source of their food and resources, fostering a sense of responsibility and appreciation for the natural world.

For example, consider Emily and Jake, a couple living in a small apartment in New York City. After years of feeling disconnected from nature, they decided to move to a small homestead in upstate New York. They began to grow their own food, raise chickens for eggs, and experiment with composting. The process of learning how to be self-sufficient and embrace the rhythms of nature provided them with not only food security but also a profound sense of satisfaction and peace.

2.   Desire for a Simpler Life

The modern world often feels chaotic and overwhelming. From the constant bombardment of digital distractions to the pressure of maintaining a high-paying job to afford an ever-growing lifestyle, many are seeking ways to escape the rat race. Minimalist homesteading allows people to simplify their lives by reducing the excess, be it physical clutter, financial burdens, or social expectations. It's about creating a life that feels purposeful, grounded, and fulfilling.

Take the example of Max and Sarah, who spent most of their early adulthood working demanding corporate jobs. After years of stress and constant busyness, they decided to downsize their home and live on a piece of land where they could grow their own food, raise animals, and focus on personal well-being. This transition allowed them to reduce their financial obligations, reduce their consumption, and spend more time focusing on what truly mattered, family, health, and creativity. The freedom they gained from embracing minimalist homesteading transformed their life.

3.   Self-Sufficiency and Sustainability

As people become increasingly aware of the environmental and economic impacts of mass consumption, many are drawn to the idea of self-sufficiency. Minimalist homesteading allows individuals to take control over their food, water, energy, and other resources. Growing one's own food, raising livestock, and reducing reliance on external sources of power not only decreases the environmental footprint but also reduces one's dependence on the global market.

In this regard, minimalist homesteading offers a lifestyle that emphasizes sustainability, using renewable resources, minimizing waste, and living within one's means. For instance, the practice of solar power on homesteads, composting waste, or rainwater harvesting reduces costs and minimizes environmental impact. This is evident in the life of John, a retired engineer who decided to downsize his home and move to a homestead in the countryside. He invested in solar panels, planted fruit trees, and built an energy-efficient home. By doing so, he was able to live off-grid, drastically reducing his utility bills while contributing to a sustainable lifestyle.

The Difference Between Minimalism and Deprivation

One of the most common misconceptions about minimalism is that it involves deprivation, living without, for the sake of simply having less. However, minimalism is not about denial; it's about intentionality and removing excess to focus on what truly matters.

1.   Minimalism as Intentionality

At its core, minimalism is about living intentionally. This means making conscious decisions about what to keep, what to eliminate, and how to live a life that is aligned with one's values. Instead of mindlessly accumulating possessions, minimalism encourages people to reflect on what is truly necessary for their well-being and happiness.

For instance, an individual who adopts a minimalist mindset might decide to downsize their wardrobe, not because they want to live without clothes, but because they want to simplify their choices and only keep the pieces that bring them joy or serve a specific purpose. This intentionality is not deprivation, but rather a way of living with more clarity and purpose.

2.   Minimalism Does Not Mean Living Without

Minimalism is often misunderstood as a form of poverty or deprivation. However, it's not about being poor or lacking the comforts of life. Instead, it's about focusing on quality over quantity. Minimalism encourages individuals to invest in fewer, higher-quality items that serve multiple purposes, rather than accumulating possessions for the sake of having more.

Take the example of Sarah, who downsized her home and sold most of her belongings. She didn't do so because she had to, but because she wanted to create a more meaningful and organized life. She invested in a few high-quality pieces of furniture that served her daily needs and brought her joy, but she no longer clung to excess items that didn't serve a purpose. Sarah still enjoyed the comforts of a beautiful home, but she did so in a way that aligned with her values.

3.   Minimalism and Fulfillment

Minimalism often leads to greater fulfillment, not less. By removing distractions and excess, people have more space to pursue the things that truly bring them joy—whether that's spending time with loved ones, focusing on personal passions, or connecting with nature. In this sense, minimalism actually opens up more opportunities for meaningful experiences and deeper satisfaction, rather than a life of scarcity.

Combining Modern Homesteading with Intentional Living

Intentional living involves making choices that are consciously aligned with your values and goals. When combined with homesteading, it forms a powerful framework for living a purposeful and fulfilling life.

1.   Living with Purpose

Modern homesteading is not simply about growing your own food or raising animals—it's about living with purpose. It's about aligning your day-to-day choices with your long-term vision for a sustainable, meaningful life. When intentional living is combined with homesteading, individuals can build lives that are deeply satisfying and aligned with their values.

Take, for example, the story of Jane, a mother of two who felt overwhelmed by the demands of city life. She and her husband decided to move to a small homestead in a rural area, where they could grow their own vegetables, raise chickens for eggs, and practice sustainable living. Every day, Jane and her family are intentional about how they spend their time. They prioritize activities that support their well-being, such as spending time outdoors, learning new homesteading skills, and engaging in creative projects. By living intentionally, they have created a life that is not only self-sufficient but also deeply fulfilling.

2.   Focus on Sustainability

Combining modern homesteading with intentional living allows individuals to focus on sustainability, both personally and environmentally. Sustainability isn't just about how you live; it's about creating systems that support the long-term well-being of the planet. Whether it's reducing waste, using renewable resources, or growing your own food, homesteading provides practical tools for living a sustainable lifestyle.

Sustainable living can look different for everyone, but it always involves being mindful of the earth's resources and finding ways to reduce your environmental footprint. For example, Emma and Mark decided to incorporate intentional living into their homesteading practices by growing organic vegetables, raising bees for honey, and using composting toilets. They also made an effort to reduce their reliance on plastic and packaging by buying items in bulk and using reusable containers. Their commitment to sustainability was not just about environmentalism—it was about creating a lifestyle that was in harmony with nature.

3.   Cultivating Mindfulness

Intentional living also encourages mindfulness—being present and fully engaged in the tasks at hand. Whether you're planting a garden, building a compost pile, or cooking a homemade meal, homesteading offers plenty of opportunities for mindfulness. In an era where technology and distractions often pull our attention in multiple directions, modern homesteading provides a way to slow down and be more present in our daily lives.

The practice of mindfulness is central to homesteading. It allows individuals to focus on the task at hand, appreciating the process of growing, building, and creating. For example, when Alex and Rachel started their homestead, they initially struggled with the many tasks they had to learn. But over time, they began to appreciate the process—whether it was learning how to care for their animals or tending to their garden. The daily work became an opportunity for mindfulness, where they could focus on each step and enjoy the rhythm of their routine.

4.   Building Community

A significant aspect of combining modern homesteading with intentional living is the building of community. Homesteading doesn't have to be a solitary endeavor. Many homesteaders actively seek out communities of like-minded individuals who value sustainability, self-sufficiency, and intentional living. By sharing knowledge, resources, and experiences, homesteaders can build supportive networks that enrich their lives and contribute to their sense of belonging.

Take the example of Greg and Rachel, who moved to a rural area and decided to build a cooperative garden with their neighbors. Together, they grew produce, raised animals, and shared tools and knowledge. The community aspect of homesteading brought them a sense of connection that enriched their lives and helped them feel supported in their journey.

In this chapter, we've explored why more people are choosing minimalist homesteading, the difference between minimalism and deprivation, and how combining modern homesteading with intentional living can create a life that is purposeful, sustainable, and fulfilling. Minimalist homesteading is not just a trend; it is a powerful lifestyle shift that allows individuals to take control of their lives, reconnect with nature, and live in alignment with their values. By embracing intentionality, sustainability, and mindfulness, people can build a life that is not only self-sufficient but also deeply meaningful.

CHAPTER 2

Designing Your Simple Homestead

Designing your simple homestead is an exciting journey that requires thoughtful planning, careful consideration, and a deep understanding of your values and goals. Whether you're dreaming of living off the grid or setting up a modest, self-sufficient home, creating the right space to nurture your family, community, and the environment is essential. This chapter will guide you through the key decisions you'll need to make when designing your homestead. From choosing the right piece of land to selecting the ideal dwelling, and from integrating energy solutions to creating spaces that truly serve your needs, the process of designing your homestead is about much more than aesthetics—it's about aligning your environment with your values and simplifying your life.

Choosing Land: What Matters and What Doesn't

The foundation of any homestead begins with the land. The right piece of land will not only shape the practicality of your homesteading lifestyle but also influence your long-term happiness and success. When searching for land, there are key factors to consider, as well as less critical elements that can be overlooked. The goal is to find a space that supports your self-sufficiency goals, aligns with your values, and fits within your budget.

1.   Location, Location, Location

The first question to ask when considering land for your homestead is: *Where do I want to live?* This decision can be influenced by factors like climate, proximity to family and friends, work opportunities, and local resources. Different areas have varying benefits and challenges, so it's essential to assess what fits best with your lifestyle.

For example, a homestead in the Pacific Northwest may provide abundant rainfall and fertile soil for gardening, but it might also come with the challenge of dealing with rainy, cold winters. On the other hand, living in the desert or the Southwest might require a greater investment in water management, but the warmer climate and drier conditions could suit those looking for a simpler, lower-maintenance living situation.

2.   Access to Resources

Your homestead's location will greatly impact your access to resources such as water, soil quality, and renewable energy options. Having access to a reliable water source (whether that's well water, a nearby stream, or a rainwater catchment system) is critical for farming, gardening, and even basic needs like drinking and sanitation. Soil quality is equally important, as fertile land can drastically reduce the amount of effort you'll need to put into growing food.

For example, on Sarah and Jack's homestead in Vermont, they chose land near a freshwater stream and invested in a natural filtration system for clean drinking water. Their property also had fertile soil, allowing them to grow a variety of vegetables without heavy reliance on chemical fertilizers. Choosing land with access to these resources allowed them to focus on homesteading practices rather than spending excessive time and money on water and soil management.

3.   Topography and Terrain

The physical layout of the land—whether it's flat, hilly, or forested—affects everything from building placement to how you plan to grow your crops. Flat land is ideal for farming, but hills and slopes may offer better drainage, making them perfect for certain crops like grapes or certain fruits. Forested land provides privacy, shelter, and the potential for sustainable timber harvesting, but it may require extra work to clear space for gardens and buildings.

For instance, Mark and Emily's land in Kentucky had a mix of forests and open fields. They chose to build their home near the edge of the forest, utilizing the natural shelter for energy efficiency and privacy, while dedicating the open fields for farming. The combination of both terrain types allowed them to adapt their homestead to both food production and comfortable living.

4.   Zoning, Regulations, and Permits

Before purchasing land, it's crucial to understand local zoning laws and regulations. Certain areas may have restrictions on building codes, animal husbandry, or land use that could impact your homesteading goals. Be sure to research the rules in your desired location regarding septic systems, water rights, waste disposal, and other key elements of sustainable living.

While researching, some prospective homesteaders discover that what initially seems like the ideal plot of land is restricted in ways that don't fit their plans. This was the case for Clara and Ben, who found a beautiful plot in their desired location but were ultimately unable to build the off-grid cabin they envisioned due to local zoning laws. They learned the importance of checking zoning ordinances early on in their search for the perfect homesteading land.

Off-Grid vs. On-Grid Light

One of the most important decisions in designing your homestead is whether to live off-grid or on-grid. Off-grid living means that you are self-sufficient in terms of water, energy, and waste disposal, while on-grid living means you are connected to municipal systems for utilities. The term "on-grid light" refers to a hybrid approach, where you're still connected to utilities like electricity or water but with the option of minimizing reliance on these systems.

1. Off-Grid Living: Pros and Cons

Off-grid living appeals to those who wish to live completely independently and minimize their environmental impact. When you live off-grid, you are responsible for generating your own power (typically using solar panels or wind turbines), collecting your own water, and managing your waste (via composting toilets or septic systems). This lifestyle provides ultimate freedom but comes with challenges—such as the upfront cost of equipment, maintenance, and the physical labor required to live self-sufficiently.

For example, James and Lisa's homestead in the mountains of Colorado is completely off-grid. They rely on solar panels for electricity, a rainwater collection system for water, and a composting toilet for waste management. While they enjoy the freedom and sustainability that comes with off-grid living, they have invested significant time and effort into maintaining their systems, especially during the long winters when sunlight is limited.

2.   On-Grid Light: A Hybrid Approach

For those who want the self-sufficiency of off-grid living but don't want to completely sever ties with modern conveniences, an on-grid light approach may be the solution. With this approach, you can still enjoy the comforts of electricity, heating, and plumbing, but you can also integrate sustainable practices like solar energy, rainwater collection, and composting.

Consider the example of Rachel and Paul, who live in a rural area on the outskirts of a small town. They have a solar panel system that supplies about 80% of their electricity needs, with the remainder drawn from the local utility grid. They also have a rainwater collection system that provides water for their garden, reducing their dependence on municipal water sources. This hybrid system allows them to enjoy the best of both worlds—maintaining a degree of self-sufficiency while still benefiting from modern infrastructure.

Tiny Homes, Cabins, and Modest Dwellings

One of the most attractive aspects of minimalist homesteading is the ability to downsize your living space. Tiny homes, cabins, and modest dwellings provide an affordable, sustainable alternative to traditional homes, which are often large, energy-inefficient, and costly to maintain. These smaller homes are designed to maximize functionality while minimizing unnecessary space.

1.   Tiny Homes

Tiny homes have become a popular choice for those looking to simplify their lives and reduce their environmental impact. With sizes typically ranging from 100 to 400 square feet, tiny homes are designed to make the most of every inch of space. They allow for lower building costs, reduced energy consumption, and less overall maintenance.

For example, David and Zoe built a tiny home on their homestead in northern California. The small footprint of their home made it easier for them to establish a garden, care for animals, and reduce their energy consumption. Despite the small size, their home was thoughtfully designed, featuring a sleeping loft, multifunctional furniture, and efficient use of storage. The tiny home allowed them to live simply while still having all the comforts they needed.

2.   Cabins and Modest Dwellings

Cabins and modest dwellings can provide more space than a tiny home while still embracing the principles of simplicity. Many homesteaders choose to build small, off-the-grid cabins that blend into the natural surroundings and offer functional living spaces. Cabins can be particularly appealing for those who want to live in more remote locations and desire a more rustic, cozy living space.

The homestead of Mark and Alice is a perfect example of a cabin that serves both as a practical and comfortable dwelling. Built with local timber, their cabin has an open floor plan with space for cooking, eating, sleeping, and working. While their cabin is more spacious than a tiny home, it remains modest and focused on functionality, ensuring they can live simply without unnecessary luxuries.

Creating Spaces That Serve, Not Stress

One of the fundamental principles of minimalist homesteading is the concept of creating spaces that serve a purpose rather than adding stress. This means designing your home, garden, and other spaces in a way that supports your homesteading goals, reduces clutter, and enhances your well-being.

1. Designing Efficiently

When designing your homestead, it's important to prioritize efficiency. Every space in your home should have a specific function, and each item you own should serve a purpose. For example, your kitchen might include ample counter space for preparing and preserving food, your garden space should be organized for easy access to planting and harvesting, and your living areas should be designed for comfort and simplicity.

2.   Simplicity in Design

Creating spaces that serve, not stress, also means embracing simplicity in design. Minimalist interiors with clean lines, neutral colors, and uncluttered spaces can reduce mental stress and create a calming environment. Thoughtful design can help you focus on the tasks that matter most, such as caring for your animals, growing food, or spending quality time with family.

Conclusion

Designing your simple homestead is a deeply personal process that involves aligning your space with your values. Whether choosing the right piece of land, deciding between off-grid and on-grid options, or selecting the ideal dwelling, your decisions will directly impact your daily life and homesteading success. By focusing on creating spaces that serve your needs and reduce stress, you can cultivate a homestead that provides both functional and emotional fulfillment, allowing you to live a simpler, more intentional life.

CHAPTER 3

The Minimalist Mindset Shift

L iving a minimalist lifestyle isn't just about decluttering your home or reducing your material possessions; it's about adopting a mindset that shifts how you approach all aspects of life. Minimalism is not just a physical practice but a mental one, too. In this chapter, we will explore how embracing a minimalist mindset can lead to a more intentional, fulfilling life. This shift involves decluttering both mental and physical spaces, prioritizing needs over wants, and letting go of consumerism while still valuing quality. By changing how we think about what we need and how we engage with the world, we can cultivate a life that is more centered, meaningful, and aligned with our deepest values.

Decluttering Mental and Physical Space

When we think of minimalism, most people focus on physical possessions—cleaning out closets, getting rid of furniture, or downsizing homes. While these are certainly important aspects, minimalism begins in the mind. In fact, decluttering your mental space can have a profound impact on your overall well-being and your ability to live intentionally.

1.   Mental Decluttering: Clearing the Noise

Mental clutter is the constant barrage of thoughts, worries, and distractions that prevent us from being present in the moment. Just as our homes can become filled with unnecessary objects, our minds can become overrun with unproductive thoughts, unresolved issues, and distractions from the digital world. The minimalist mindset encourages us to clear this mental clutter by focusing on what is truly important and letting go of things that do not serve our overall well-being.

For example, many people find themselves overwhelmed by endless to-do lists, social media notifications, and the pressure to constantly achieve. By practicing mindfulness, setting boundaries, and intentionally choosing what we allow our minds to focus on, we can reduce mental clutter and improve our mental clarity. This allows us to make better decisions, be more present, and focus our energy on the things that truly matter.

John, a graphic designer living in a busy city, found that his mental clutter was negatively impacting his creativity. His phone notifications, work stress, and constant stream of social media updates left him feeling scattered and unfocused. To declutter his mind, John made a conscious effort to turn off non-essential notifications, set aside specific times for checking emails, and incorporate daily meditation into his routine. This shift helped him reclaim his mental space, enabling him to feel more grounded and focused on his creative work.

## 2.	Physical Decluttering: Creating Space for What Matters

When it comes to physical space, minimalism involves consciously eliminating things that do not add value to your life. Whether it's clothes you no longer wear, kitchen gadgets gathering dust, or sentimental items that no longer serve you, clearing out physical clutter can have a profound impact on both your living environment and your mental state. The process of decluttering is not about being extreme or living without, but about being intentional with what you keep.

Consider Anna, a mother of three, who found her home overwhelmed by toys, clothes, and endless piles of paperwork. The clutter was causing stress and frustration, as there was never enough time to clean, organize, or simply enjoy her home. After embracing a minimalist approach, Anna took it step by step, focusing on one room at a time. She began by donating or selling toys that no longer served her children's interests and sorting through old clothes that no longer fit. The result was a more open, organized living space that was easier to maintain. Not only did the physical space feel more peaceful, but Anna found that the mental clarity she gained from decluttering made her feel less anxious and more focused.

Incorporating minimalist principles into your home design can also enhance the function of your living space. By removing items that are not essential, you create a space that is easier to navigate, more organized, and conducive to relaxation. Each item in your home should serve a purpose, whether it's functional, aesthetic, or sentimental. The act of decluttering is about making space for what truly adds value to your life, and letting go of the things that are simply taking up space, both physically and mentally.

Nicci Brochard & Dr.Ben Chuba

Prioritizing Needs Over Wants

One of the central tenets of minimalism is the shift from focusing on wants to focusing on needs. In today's consumer-driven world, it's easy to get caught up in the idea that we need more—more clothes, more gadgets, more conveniences. The minimalist mindset challenges this notion by encouraging us to evaluate what we truly need and what we can live without.

1.   Understanding True Needs

What is the difference between a need and a want? Needs are the basic necessities that sustain life, food, shelter, water, and clothing. Wants, on the other hand, are desires that arise from social pressures, advertising, and the idea that more will bring happiness. Minimalism encourages us to examine our desires and distinguish between what we truly need to live a fulfilling life and what we want out of habit or societal expectations.

For instance, many people think they need the latest smartphone, the trendiest clothes, or the newest car model because those things are heavily marketed as symbols of success. However, a minimalist approach asks: *Do I need this new item to live well, or am I just chasing the idea of more?*

When Rachel and Tom moved to their homestead, they intentionally focused on their needs—good food, shelter, and tools for their farm—and avoided purchasing unnecessary items. For example, when Rachel wanted a new dress for a special occasion, she instead chose to buy secondhand clothes, prioritizing quality over quantity and longevity over fleeting trends. This mindset not only helped them save money but also reduced the mental and physical clutter in their lives.

2.   The Satisfaction of Simplicity

By focusing on needs, minimalism teaches us to appreciate the value of simplicity. It encourages us to use fewer resources, own fewer things, and invest in quality items that will serve us overtime. When we stop constantly seeking the next thing, we begin to realize that true satisfaction comes not from accumulation, but from cultivating a life that aligns with our core values.

A good example is Marcus, who realized that his constant desire for new gadgets and clothes was taking a toll on his finances and mental peace. After embracing minimalism, Marcus started buying fewer but higher-quality items, choosing items that truly added value to his life. Over time, he found that his satisfaction no longer stemmed from the thrill of a new purchase but from the joy of living with less, being more mindful, and making thoughtful decisions. Marcus's mindset shift toward prioritizing needs helped him foster a sense of contentment and freedom.

Letting Go of Consumerism While Still Valuing Quality

Letting go of consumerism is one of the most liberating aspects of adopting a minimalist mindset. Consumerism tells us that happiness is achieved through constant acquisition, but minimalism teaches us that satisfaction comes from living in alignment with our values and being intentional about our choices. This doesn't mean we must abandon all consumption—rather, it's about shifting our mindset from mindless acquisition to thoughtful, value-driven consumption.

1.   Understanding Consumerism

Consumerism encourages the constant pursuit of more—more products, more options, more entertainment, more social validation. It creates a cycle of desire, leading to the belief that happiness and fulfillment lie just beyond the next purchase. Minimalism, on the other hand, challenges this idea by promoting the notion that enough is enough. It invites us to embrace quality over quantity and to be more mindful of the impact our choices have on both ourselves and the planet.

For example, rather than buying new clothes every season to keep up with trends, minimalists might choose to invest in timeless, high-quality pieces that last longer. This not only saves money over time but also reduces waste and consumption. When choosing products or services, minimalists prioritize those that align with their values, whether it's supporting small businesses, buying ethically sourced items, or choosing products that have a minimal environmental impact.

2.   Valuing Quality Over Quantity

Minimalism doesn't mean rejecting all material possessions or avoiding purchases altogether. Rather, it emphasizes choosing quality over quantity. By valuing quality, we make purchases that last longer, perform better, and offer true value, rather than simply acquiring things for the sake of having more.

This shift not only leads to more thoughtful purchasing decisions but also reduces waste, which is a crucial aspect of a sustainable, minimalist lifestyle.

Take the example of Samantha, a graphic designer who used to spend money on cheap clothing and accessories that wore out quickly. After adopting a minimalist mindset, Samantha shifted her focus to buying fewer, higher-quality items that would last longer. She invested in well-made clothing, sourced from sustainable brands, and took care of her belongings to ensure they would serve her for years to come. By valuing quality over quantity, Samantha was able to cultivate a wardrobe that aligned with her values, was more versatile, and reduced the need for constant replacements.

Conclusion

The minimalist mindset is not just about decluttering your home or owning fewer things; it's about embracing a way of life that is centered on intentionality, simplicity, and value. By decluttering both our physical and mental spaces, prioritizing needs over wants, and letting go of consumerism while valuing quality, we can create a life that is not only more peaceful but more aligned with our true values. Minimalism encourages us to live with purpose, reduce distractions, and focus on what truly matters. By shifting our mindset, we begin to experience a deeper sense of fulfillment, freedom, and clarity—ultimately creating a life that is not weighed down by the excess but uplifted by the essentials.

CHAPTER 4

Powering Down – Simple Energy Solutions

As we move towards a more sustainable lifestyle, one of the most significant shifts we can make is in how we power our homes and daily lives. Minimalist homesteading and intentional living are deeply connected to reducing our reliance on the grid and embracing alternative energy solutions that are not only eco-friendly but also cost-effective. This chapter will explore simple energy solutions for homesteaders, focusing on solar, wind, and micro-hydro power systems, reducing dependence on traditional power grids, choosing energy-conscious appliances, and harnessing the power of natural light and seasonal rhythms. Each of these solutions can be an essential part of living more sustainably and in harmony with the environment.

Solar, Wind, and Micro-Hydro Basics

The three primary sources of renewable energy for off-grid and eco-conscious living are solar, wind, and micro-hydro. These energy systems allow homesteaders to take control of their power needs, reduce utility bills, and minimize their environmental footprint.

1.   Solar Power

Solar power is perhaps the most accessible and widely used form of renewable energy. It involves capturing sunlight with photovoltaic (PV) panels, which convert sunlight into electricity. Solar power is particularly ideal for homesteaders who live in sunny regions, as it provides an almost unlimited supply of energy during the day.

How Solar Power Works

Solar panels are installed on the roof or a dedicated structure where they can capture maximum sunlight. The electricity generated by the panels is either stored in batteries for later use or directly routed to power the home. A solar power system typically includes an inverter, which converts the DC power generated by the solar panels into AC power used by most household appliances.

For example, the Smith family, living on a small farm in Arizona, installed a 5-kilowatt solar system. They chose solar because of the high number of sunny days in their area and their desire to reduce dependence on the grid. The solar panels cover about 75% of their electricity needs, and they have a backup generator to meet higher demand during cloudy days or at night. Solar power has helped them save significantly on energy costs while reducing their carbon footprint.

2.   Wind Power

Wind power is another excellent option for homesteaders, particularly those living in areas with consistent wind patterns.

Wind turbines convert kinetic energy from the wind into electrical energy. A wind turbine is most effective in open, windy areas where the wind speed is consistently above 9 mph.

How Wind Power Works

A wind turbine consists of large blades connected to a rotor, which spins when the wind blows. The rotor's movement generates electricity, which can be stored in batteries or used directly to power appliances. Small-scale wind turbines designed for residential use are typically around 20 to 30 feet tall, though the size and type of turbine depend on the wind conditions in the area.

For instance, the Johnsons, who live in a rural part of the Midwest, installed a small wind turbine on their homestead. Their farm is located on an open plain, where wind speeds average 15 mph, making it an ideal location for wind power. Their turbine supplements their solar power system, ensuring a steady and reliable energy supply even during the winter months when sunlight is limited. Together, these renewable systems have allowed the Johnsons to reduce their energy costs and become less reliant on traditional grid power.

3.   Micro-Hydro Power

Micro-hydro power systems are ideal for homesteaders who live near a river or stream. These systems harness the power of moving water to generate electricity. Micro-hydro systems can produce a significant amount of energy, even more than solar or wind systems, provided there is a reliable and consistent water source.

How micro-Hydro Power Works

Micro-hydro systems use the flow of water to turn a turbine, which generates electricity. The water flow turns the turbine, which powers a generator. The energy is then either stored in batteries or used directly to power the home. The key to micro-hydro systems is having a constant and reliable water source, such as a stream or river with sufficient flow and drop (or "head") to generate power.

Take the example of Anna and Luke, who live on a homestead nestled in the Appalachian Mountains. They have a small stream that runs through their property, which they've used to power a micro-hydro system. This system generates electricity year-round, even during the darkest days of winter when solar power is less effective. Their micro-hydro system powers everything from lights and appliances to their farm equipment, reducing their need for external power sources and providing a nearly continuous supply of energy.

Reducing Dependence on the Grid

One of the main goals of minimalist homesteading and sustainable living is reducing or eliminating reliance on traditional power grids. Living off the grid allows homesteaders to be more self-sufficient, lower their utility bills, and reduce their environmental impact. However, reducing grid dependence requires thoughtful planning and investment in energy-efficient systems.

1.   Energy Storage and Backup Systems

While solar, wind, and micro-hydro systems provide renewable energy, they often require storage solutions to ensure a constant power supply. Batteries allow energy produced during the day to be stored for use at night or during periods of low generation (e.g., cloudy days for solar or calm periods for wind).

For example, Chris and Martha, who live in a remote area of Alaska, rely on a combination of solar panels and a small wind turbine to meet their energy needs. They installed a battery storage system to ensure they have enough power during the long winters when sunlight is scarce, and winds are not always consistent. Having a backup power system means that they can live without worrying about power outages or disruptions to their energy supply.

2.   Energy-Efficient Homes

Designing a home that requires minimal energy to heat, cool, and power is another key step in reducing grid dependence. Energy-efficient homes use materials and design strategies that minimize heat loss in the winter and reduce the need for air conditioning in the summer. For example, insulated windows, thick walls, and passive solar heating can significantly reduce the need for external power to maintain a comfortable living environment.

Sarah and Ben, who recently built their home on their homestead, made sure to incorporate passive design techniques, such as large south-facing windows for solar heating, a thermal mass wall to store and release heat, and a high level of insulation. These features allow their home to stay warm in the winter and cool in the summer without relying on electric heating or air conditioning.

3.   Smart Power Management

Power management systems allow homesteaders to track and optimize their energy use, making it easier to reduce waste and ensure energy-efficient operation. Many systems now offer smart meters that allow users to monitor their energy consumption in real-time and adjust accordingly. By knowing how much power is being used, homesteaders can make informed decisions about when to use certain appliances or systems to conserve energy.

Take the example of Derek and Emma, who live in an off-grid cabin. They installed a smart energy management system that allows them to monitor their solar, wind, and micro-hydro systems in real-time. This system helps them track energy generation and consumption, enabling them to adjust usage when necessary and ensure that their battery storage is sufficient for their needs.

Energy-Conscious Appliances and Tools

While generating your own energy is essential for living sustainably, it's also important to use energy-efficient appliances and tools that reduce overall consumption. By prioritizing energy-efficient devices, homesteaders can lower their energy demand, reduce their reliance on external power, and minimize their environmental footprint.

1.   Energy-Efficient Appliances

From refrigerators to washing machines, many modern appliances now come with energy-efficient features that use less power while still providing excellent performance. Energy Star ratings are a good indicator of an appliance's efficiency, as these products are designed to consume less electricity while providing the same level of functionality.

For example, Jessica and Rob have an off-grid homestead in the mountains, where they rely on solar power to run their home. They installed an energy-efficient refrigerator, washing machine, and water pump to minimize energy use. By choosing appliances with high energy ratings, they've been able to reduce their overall energy consumption, allowing them to make the most of their renewable energy systems.

2.   Efficient Cooking and Heating

Cooking and heating often account for a significant portion of energy usage in a home. By choosing more efficient cooking methods, such as a wood stove or an induction cooktop, homesteaders can reduce their reliance on grid-based electricity. Wood stoves, for example, provide both heat and a way to cook, making them a valuable tool for off-grid living.

Similarly, choosing an energy-efficient heating system, like a wood-burning stove or passive solar heating, can dramatically reduce the need for electricity-based heating.

Embracing Natural Light and Seasonal Rhythms

One of the simplest ways to reduce energy consumption is by embracing natural light and aligning your lifestyle with the rhythms of the seasons. By designing your homestead to take full advantage of natural light, and by living in tune with seasonal patterns, you can minimize the need for artificial lighting and climate control.

1.   Maximizing Natural Light

Natural light not only reduces the need for electric lighting but also contributes to a healthier living environment. Properly positioned windows, skylights, and light-colored interiors can help capture and distribute sunlight throughout the day.

For example, Amanda and Tom built a passive solar home with large windows positioned to take advantage of sunlight throughout the day. They also included skylights in their bedrooms and living areas to maximize natural light, making their home feel bright and warm without relying heavily on electricity.

2.   Living in Tune with Seasonal Rhythms

Embracing seasonal rhythms means adjusting your daily routines to align with the natural ebb and flow of the seasons. In winter, you may spend more time indoors, using less energy as you huddle by the fire, while in the summer, you may rely more on natural cooling methods, such as opening windows to catch a breeze.

For instance, during the long summer days, Daniel and Susan, who live on a small homestead in the Pacific Northwest, use natural cooling methods, such as cross-ventilation and shading their windows, to keep their home comfortable without relying on air conditioning. In winter, they embrace the slower pace of life, enjoying cozy evenings by the wood stove, where they use less energy for heating while staying warm and content.

Conclusion

Incorporating simple energy solutions into your homestead can drastically reduce your reliance on the grid, lower your energy costs, and help you live in harmony with the environment. Whether you choose solar, wind, or micro-hydro power systems, embrace energy-efficient appliances, or design your home to harness natural light, these choices are key to creating a sustainable and self-sufficient lifestyle. By reducing our dependence on external energy sources and aligning with natural rhythms, we can cultivate a life that is not only more cost-effective but also more fulfilling and eco-conscious.

CHAPTER 5

Water Wisdom – Collection, Conservation, and Clean Use

Water is the essence of life. It sustains plants, animals, and humans alike. However, as global populations grow, water resources are becoming increasingly strained. In many regions, particularly those dependent on agriculture or living off-grid, managing water efficiently has become more important than ever. Minimalist homesteading embraces the idea of sustainability, and one of the cornerstones of sustainability is water conservation and management. This chapter will dive deep into the concepts of rainwater harvesting, greywater reuse, composting toilets, and how to live with less water, while ensuring that the water you use is clean, safe, and effectively managed.

Rainwater Harvesting and Filtration Systems

Rainwater harvesting is one of the most straightforward and effective ways to reduce dependency on municipal water systems while securing an independent water source. By capturing rainwater from roofs and storing it for future use, homesteaders can take advantage of a free, renewable resource that is often underutilized. However, as with any water source, it's crucial to ensure that the harvested rainwater is filtered and purified before use, especially for drinking and cooking.

1.   The Basics of Rainwater Harvesting

Rainwater harvesting begins with the collection of rainwater from rooftops. The process typically involves a system of gutters, downspouts, and storage tanks. These systems are designed to channel rainwater from your roof into a storage container, where it can be stored for later use. The key benefit of rainwater harvesting is that it takes advantage of a natural cycle, ensuring that even in areas where fresh water may be scarce, rainwater can be captured and utilized.

How It Works:

- **Catchment Area (Roof):** The roof acts as a catchment area where rainwater is collected. It's important to ensure the roof is clean and made of non-toxic materials to prevent contamination of the water.

- **Gutters and Downspouts:** The water is directed from the roof into gutters, which channel the water to downspouts leading to storage tanks. Proper maintenance of the gutters ensures that the water is filtered to some extent as it is funneled into the system.

- **Storage Tanks:** Rainwater is stored in tanks, which can be made from various materials, including plastic, steel, or concrete. The size of the tank depends on the amount of rainfall in the area and the household's water usage needs. For a homestead, large cisterns or barrels are typically   used.

- **Filtration and Treatment:** Once the rainwater is collected, it's crucial to filter it before use. Basic filtration systems include debris screens and first-flush diverters to prevent contaminants like dirt, leaves, and animal droppings from entering the tank. More advanced systems may include carbon filters, UV purification, or reverse osmosis, particularly if the water is to be used for drinking or cooking.

2.   Types of Rainwater Harvesting Systems

Rainwater harvesting systems can be classified into different types, depending on the complexity and the intended use of the water:

- **Basic System:** A simple system involves gutters and downspouts leading directly to a storage tank with a basic filtration system (like a debris screen or first-flush diverter). This type of system is ideal for non-potable uses such as irrigation or washing.

- **Advanced System:** These systems include additional filtration, such as sediment filters, carbon filters, UV sterilizers, or reverse osmosis systems. These systems are necessary for households where the water will be used for drinking and cooking.

For instance, in Jane's off-grid homestead in the desert Southwest, she installed a rainwater harvesting system that captures water from her roof. She uses a basic system for irrigation in her garden but also has a second system that includes a UV filter for drinking water. Over the course of the year, this system has reduced her need to purchase water, especially in dry months.

3.   Maintenance and Quality Control

It's essential to regularly maintain rainwater harvesting systems to ensure that the water remains clean and safe for use. The first-flush system, which diverts the initial flow of water during the first rain after a dry spell, helps to ensure that contaminants, such as bird droppings and debris, are not collected. Regular cleaning of gutters, checking for blockages in the downspouts, and inspecting storage tanks for mold or algae buildup are also necessary tasks.

For example, Mark and Lisa, who live in an area with seasonal rains, installed a first-flush system, and they clean their tanks every few months to prevent algae growth. By ensuring that the storage and filtration system is well-maintained, they can continue using the rainwater without concerns about contamination.

Greywater Reuse and Composting Toilets

Greywater reuse and composting toilets are two innovative systems that can significantly reduce the amount of fresh water needed on a homestead while promoting sustainability. Greywater refers to wastewater from sources like showers, sinks, and washing machines that can be reused for irrigation and other non-potable uses. Composting toilets are an eco-friendly way of dealing with human waste by turning it into compost that can be safely used to fertilize plants.

1.   Greywater Reuse

Greywater is the water that flows down the drain after everyday activities such as showering, washing hands, or washing clothes. It is different from blackwater, which contains human waste from toilets. Because greywater is relatively free of harmful pathogens, it can be safely reused for non-potable purposes, like irrigating your garden, flushing toilets, or washing outdoor areas. The key benefit of greywater reuse is that it reduces the demand for fresh water while also reducing the strain on municipal sewage systems.

How to Reuse Greywater

Greywater systems typically consist of a filtration system that removes debris and contaminants. Once filtered, the water is directed to a storage tank or directly into irrigation systems. There are two main types of greywater systems:

- **Simple Systems:** These systems collect water from one or two sources, such as a shower or laundry, and redirect it to the garden or yard.

- **Complex Systems:** These systems may collect greywater from multiple sources and include a more advanced filtration and treatment process to ensure the water is safe for reuse.

For instance, Claudia, living on a small homestead, installed a greywater system that collects water from her laundry and shower. The system includes a simple filtration system that filters out hair and soap residue, then diverts the water to her garden. By reusing this water, she has significantly reduced the amount of water needed to maintain her garden and landscaping.

2.   Composting Toilets

Composting toilets are a great way to manage human waste in a sustainable and eco-friendly manner. Unlike traditional flushing toilets that use large amounts of water to transport waste to a septic system or sewage plant, composting toilets use little to no water and convert human waste into compost. This not only conserves water but also produces a valuable resource for the garden.

How Composting Toilets Work

Composting toilets work by separating liquid waste from solid waste, allowing aerobic bacteria to break down the solids into compost. A ventilation system is typically used to ensure airflow, which helps the composting process. Some models include a hand crank or rotating drum to mix the waste and facilitate decomposition. Over time, the waste breaks down into nutrient-rich compost, which can be used to fertilize non-food plants, trees, or ornamental gardens.

For example, the McAllister family, who live on a homestead in the Pacific Northwest, use a composting toilet as part of their off-grid lifestyle. They found that it not only saves water but also reduces their reliance on septic systems, which can be costly and environmentally damaging. The compost they generate from their toilet is used to enrich their garden soil, closing the loop on waste and nutrients.

Living with Less: How Much Water Do We Really Need?

The average person uses 80 to 100 gallons of water a day, a significant portion of which is used for activities like flushing toilets, showering, and watering lawns. But the reality is that we can live with much less water, especially when we prioritize conservation and use water more mindfully. The minimalist mindset encourages us to question our water usage and determine how much we truly need.

1. Daily Water Consumption: Breaking Down the Numbers

To understand how much water we actually need, it's useful to break down our daily usage. A typical household might use:

- **Showering/Bathing:** 10-25 gallons per shower (depending on the type of showerhead).

- **Toilets:** 1.6-5 gallons per flush (depending on the type of toilet). **Dishwashing:** 5-15 gallons per load.

- **Laundry:** 15-40 gallons per load (for a washing machine).

- **Cooking and Drinking:** 1-2 gallons per person.

These numbers add up quickly, but with mindful adjustments, we can drastically reduce our daily consumption. For example, replacing old toilet fixtures with low-flow models can save gallons of water per day. Likewise, reducing shower time, using water-efficient appliances, and collecting rainwater for outdoor use can all contribute to living with less water.

2.   Water-Saving Tips for the Homestead

To live with less water, homesteaders can implement a number of practical strategies. These include:

- **Using Low-Flow Fixtures:** Installing low-flow showerheads, faucets, and toilets can reduce water usage without compromising comfort.

- **Reducing Shower Time:** Simple changes like taking shorter showers or using a bucket to catch water during warm-up can save gallons of water every day

- **Collecting Greywater for Outdoor Use:** Greywater can be repurposed for irrigation, reducing the need to use fresh water on your garden or landscape.

- **Watering the Garden Efficiently:** Watering the garden in the early morning or late evening reduces evaporation, and using drip irrigation systems ensures that water goes directly to the roots where it's needed most.

3.   Water Conservation in Everyday Life

Living with less water requires a shift in mindset. By embracing water conservation as a way of life, homesteaders can not only reduce their impact on the environment but also save money and resources in the long term. Simple habits like turning off the tap while brushing your teeth, fixing leaky faucets, and being mindful of water usage in the garden can all make a significant difference.

For instance, Lily, who lives on a small homestead, made the decision to reduce her water consumption by installing a low-flow showerhead, using a rainwater system for garden irrigation, and composting rather than using a flush toilet. These small steps, combined with mindful habits, have allowed her to reduce her household water usage by over 50%.

Conclusion

Water is one of our most valuable resources, and learning to manage it efficiently is crucial for anyone striving to live a minimalist, sustainable life. Rainwater harvesting, greywater reuse, and composting toilets provide simple yet effective ways to conserve water while maintaining the comfort and functionality of your homestead. At the same time, embracing the minimalist mindset by questioning how much water we truly need and making small changes in our daily routines can have a profound impact on our water consumption. Ultimately, water wisdom—collection, conservation, and clean use— empowers us to live more sustainably, make the most of the resources available to us, and contribute to a healthier planet.

CHAPTER 6

Growing with Intention – The Minimalist Garden

In the pursuit of minimalist homesteading, the garden becomes a vital space for both sustainability and personal fulfillment. Gardening allows us to reconnect with the earth, cultivate our own food, and minimize our reliance on external resources. However, creating a minimalist garden is not about maximizing space or growing an abundance of crops—it's about growing with intention. Whether you have a small urban garden, a sprawling homestead, or an apartment balcony, the principles of a minimalist garden can be adapted to any space. In this chapter, we will explore how to start small and seasonal, utilize efficient gardening techniques like raised beds, permaculture, and container gardening, and incorporate high-efficiency planting and food preservation practices. We will also discuss how composting can be a cornerstone of a minimalist garden, enriching the soil while reducing waste.

Starting Small and Seasonal

When it comes to creating a minimalist garden, the key is to start small. Trying to plant an entire garden all at once, especially as a beginner, can lead to overwhelm and frustration. Minimalism teaches us to focus on quality over quantity, and this applies to gardening as well. Starting small and seasonal allows you to focus on the crops that will thrive in your climate and time frame, reducing waste, and ensuring that your gardening efforts are both manageable and productive.

1.   Start with What You Need

The first step in creating a minimalist garden is determining what you actually need and use in your day-to-day life. A minimalist garden is about providing yourself with food that is essential, practical, and meaningful. There's no need to grow crops that you rarely eat or that take up excessive space in your garden. Instead, choose plants that will enhance your meals and contribute to a simple, nutritious lifestyle.

For instance, if you enjoy cooking with fresh herbs, then growing a small herb garden with basil, rosemary, thyme, and mint would be a great choice. If you use a lot of tomatoes in your cooking, a few tomato plants may provide all the fresh tomatoes you need, without overwhelming your space or taking up too much of your gardening time.

Take the example of Carla, who started a minimalist garden in her backyard. Instead of planting every vegetable under the sun, she carefully selected crops like spinach, kale, tomatoes, and bell peppers, which were staples in her diet. She found that by focusing on a few, well-chosen plants, she could manage her garden more efficiently and enjoy fresh produce throughout the growing season without the stress of overplanting.

2.   Plant Seasonally

Minimalist gardening also involves understanding the natural rhythms of the growing season. By planting seasonally, you can avoid the overuse of resources such as water and fertilizer. Planting seasonal crops also ensures that you're working in sync with the environment, maximizing the yields and minimizing the effort.

Seasonal gardening focuses on growing the right crops at the right time. For example, cool-season crops like lettuce, peas, and spinach are perfect for early spring or fall, while warmer-season crops like tomatoes, peppers, and beans thrive in the summer months. By planting according to the season, you can make the most of your growing space and effort, ensuring that your garden is productive without becoming overwhelming.

Sam and Lisa, a couple living on a small homestead, practice seasonal gardening by focusing on cold-hardy crops in the early spring and late fall, while growing warm-weather crops like squash and cucumbers in the summer. They have learned to adapt to the seasonal cycle, ensuring that they always have fresh produce on hand without overburdening themselves with maintenance.

Raised Beds, Permaculture, and Container Gardening

In minimalist gardening, space efficiency is key. Raised beds, permaculture techniques, and container gardening are excellent ways to make the most of limited space while maximizing plant yields. These methods also align with minimalist principles by reducing waste and focusing on systems that are sustainable and low-maintenance.

1.   Raised Beds: Maximizing Space and Soil Quality

Raised beds are a highly effective and popular gardening technique, particularly for small spaces or areas with poor soil quality. A raised bed is essentially a garden plot that is elevated above the surrounding ground, typically contained by a frame made of wood, stone, or metal. The soil in a raised bed is often richer and more fertile than in ground gardens, which is ideal for growing high-yield crops like vegetables, herbs, and even flowers.

Benefits of Raised Beds:

- **Better Drainage:** Raised beds drain more easily, reducing the risk of waterlogging and root rot.

- **Improved Soil Quality:** You can customize the soil mix in a raised bed to ensure that it's rich in nutrients and free from contaminants.

- **Space Efficiency:** Raised beds allow for closer spacing of plants, maximizing the growing area without overcrowding.

- **Reduced Maintenance:** Raised beds can be easier to weed and maintain since the soil is more contained and accessible.

For example, John and Anna, who live in a small suburban backyard, built several raised garden beds to grow their vegetables. By focusing on a small, highly productive space, they were able to grow enough produce for their family of four while keeping their garden manageable. Raised beds allowed them to avoid the struggle with their clay-heavy soil, and the garden required less maintenance compared to traditional in-ground plots.

2.   Permaculture: Designing for Sustainability

Permaculture is a holistic approach to gardening that seeks to design ecosystems that mimic natural processes. It emphasizes sustainability, diversity, and minimal intervention. By working with nature rather than against it, permaculture allows you to create a garden that requires less input, reduces waste, and increases biodiversity.

Key Principles of Permaculture:

- **Diversity and Polyculture:** Rather than growing a single crop in a large area, permaculture encourages planting a variety of crops that support each other. Companion planting, where certain plants benefit from being grown together, is a common practice in permaculture gardens.

- **Natural Systems:** Permaculture designs use natural processes, such as water cycling and plant interactions, to minimize external inputs. For example, planting nitrogen-fixing plants like beans alongside other crops reduces the need for synthetic fertilizers.

- **Resource Efficiency:** Permaculture systems focus on using resources efficiently, such as collecting rainwater, composting waste, and utilizing natural mulch to preserve soil moisture.

For instance, the Cooper family on their 3-acre homestead in Oregon embraced permaculture principles by creating swales to capture rainwater, planting a food forest that mimicked the natural ecosystem, and using chickens to help with pest control. Their garden flourished with minimal effort, and the diversity of plants supported each other in a mutually beneficial system.

3.   Container Gardening: Flexibility and Portability

For those with very limited space, container gardening provides a perfect solution. Container gardening involves growing plants in pots, tubs, or other containers rather than in the ground. This technique is highly flexible, as it allows you to grow plants on patios, balconies, or even windowsills.

Benefits of Container Gardening:

- **Portability:** Containers can be moved to take advantage of sun, shade, or shelter from the wind.

- **Space-Saving:** Containers allow you to grow a variety of plants in even the smallest of spaces.

- **Easy Access:** Container gardens are often easier to maintain, as they reduce the need for bending or kneeling, which is particularly useful for those with limited mobility.

For example, Lily, who lives in an urban apartment, uses containers to grow tomatoes, peppers, and herbs on her balcony. She uses large pots and self-watering containers to ensure the plants stay hydrated, even when she's away for a few days. By growing plants in containers, she's able to enjoy fresh food without requiring much space or effort.

High-Efficiency Planting and Food Preservation

One of the key benefits of minimalist gardening is that it emphasizes efficiency. High-efficiency planting and food preservation techniques allow homesteaders to make the most of their garden space and food production while minimizing waste.

1. High-Efficiency Planting

High-efficiency planting focuses on maximizing yield while minimizing the amount of space, water, and energy used. Techniques like square-foot gardening, intensive planting, and succession planting are perfect examples of how to get the most out of a small garden.

Square-Foot Gardening: This method involves dividing the garden into small square sections (usually 1-foot by 1-foot) and planting a specific number of plants in each square based on the plant's size. This method maximizes plant density and minimizes wasted space.

Succession Planting: Succession planting involves planting crops in succession rather than all at once. By planting crops in stages, you can extend the harvest season and make the most of your growing space. For example, after harvesting early-season crops like spinach, you can plant late-season crops like kale or carrots in the same space.

2.   Food Preservation

Once you've grown your garden, preserving the harvest is an essential step in minimalist homesteading. Canning, freezing, drying, and fermenting are all methods of preserving food that allow you to enjoy your produce long after the growing season is over. Food preservation reduces waste and ensures that you're able to make the most of your harvest, even when fresh produce isn't available.

For instance, when Sarah harvested her tomatoes, she canned them for use in soups, stews, and sauces throughout the year. She also dried herbs from her garden and preserved them in jars, using them in her cooking long after the growing season ended.

Composting for the Minimalist

Composting is a cornerstone of minimalist gardening. Not only does it reduce waste, but it also improves soil fertility, enhances plant health, and promotes sustainability. Composting allows you to recycle organic waste into nutrient-rich soil that can be used to nourish your garden.

1.   Composting Basics

Composting involves breaking down organic material such as food scraps, yard waste, and plant matter into nutrient-rich compost that can be used to fertilize plants. A well-maintained compost bin or pile can decompose organic material in a few months, depending on the temperature, moisture, and type of waste.

For example, the Turner family maintains a simple compost bin in their backyard where they add fruit and vegetable scraps, leaves, and grass clippings. They turn the compost regularly to speed up the decomposition process, and in about six months, they have a rich, dark compost that they use to fertilize their vegetable garden.

3.   Composting Toilets

For the minimalist homesteader, composting toilets are a valuable tool for reducing water use and minimizing waste. Rather than relying on flush toilets that use large amounts of water, composting toilets separate waste into solids and liquids, allowing the solids to break down into compost over time. This compost can be used to enrich the soil, though it should be done so with care to avoid contamination of edible crops.

For example, Alice and Jake have a composting toilet in their off-grid cabin in the mountains. They use a simple composting system that collects waste in a bin, where it is mixed with sawdust to help with decomposition. After a year, the compost is used to fertilize non-food plants in their yard, completing the cycle of waste to resources.

Conclusion

The minimalist garden is a space for both practicality and mindfulness. By starting small and seasonal, utilizing raised beds, permaculture, and container gardening, and embracing high-efficiency planting, minimalist homesteaders can cultivate food that nourishes their bodies while also promoting sustainability. Additionally, by composting and practicing food preservation, they can ensure that nothing goes to waste, further reducing their reliance on external resources. Gardening with intention is not just about growing food—it's about fostering a deeper connection to nature, reducing environmental impact, and living a life that aligns with the principles of minimalism. Through thoughtful planning and mindful practices, a minimalist garden becomes an essential tool for self-sufficiency and sustainability.

CHAPTER 7

Animals and the Simple Life

In minimalist homesteading, animals can play a pivotal role in creating a sustainable, self-sufficient lifestyle. Whether you're looking to raise chickens for fresh eggs, goats for milk, or bees for honey, animals provide essential resources that can support your homesteading efforts. However, the minimalist mindset encourages us to keep only what we truly need, ensuring that the animals we raise align with our goals and contribute to the sustainability of the homestead without causing unnecessary stress or burden. This chapter will delve into how to raise animals in a minimalist, ethical, and sustainable way, focusing on animals like chickens, goats, and bees, while also discussing their feed, shelter, and waste management in small-scale systems.

Chickens, Goats, Bees, and Beyond: Keeping Only What You Need

When it comes to raising animals on a homestead, the minimalist approach is all about keeping the number of animals manageable, choosing species that are well-suited to your environment, and ensuring that each animal plays a functional role in your homesteading goals. The idea is not to collect animals for the sake of having them, but rather to raise them for specific needs; whether for food, fiber, or other resources.

1.   Chickens: Eggs and Meat for the Homestead

Chickens are often the first animal many homesteaders choose to raise due to their relatively low maintenance and high productivity. A small flock of chickens can provide fresh eggs daily, which is invaluable for anyone seeking self-sufficiency. Additionally, chickens can be raised for meat, adding to your food security without the need for large-scale farming.

Considerations for Raising Chickens:

- **How Many Do You Need?** For a small family, 3 to 6 hens will typically be enough to provide fresh eggs. Too many chickens can lead to overcrowding and unnecessary waste. It's important to keep your flock size manageable based on your family's consumption.

- **Breed Selection:** Different chicken breeds offer different advantages. For example, the Rhode Island Red is known for being a prolific egg layer, while the Cornish Cross is commonly raised for meat. Choose breeds that align with your needs—whether for eggs, meat, or even ornamental purposes.

- **Coop Space:** Chickens need a secure coop to protect them from predators. However, minimalist homesteading doesn't mean building an elaborate, resource-heavy coop. A simple, functional chicken coop made with recycled materials can provide enough space for your chickens to thrive.

For instance, Claire and John, a couple living in the foothills of North Carolina, started with just four hens. They built a small, basic coop using reclaimed wood and wire mesh, providing their hens with enough space to roam and a safe place to sleep at night. Their chickens produced enough eggs to supply their family, and any surplus was sold at a local farmers' market. By keeping their flock small and the coop simple, Claire and John found the balance between providing for their needs and minimizing unnecessary work or resources.

2.   Goats: Milk and Fiber with Minimal Effort

Goats are a wonderful addition to any minimalist homestead, especially if you want fresh milk, cheese, or fiber (from breeds like Angora or Cashmere goats). Goats are relatively low-maintenance compared to larger animals like cows, and they are perfect for small-scale homesteading. They can also help with land management by grazing on weeds, grass, and brush.

Considerations for Raising Goats:

- **How Many Do You Need?** One or two goats may be sufficient for a small homestead. A single goat can provide about one to two quarts of milk a day, which may be enough for a small family. If you want milk for multiple households or for cheesemaking, you might consider adding more goats.

- **Choosing the Right Breed:** Different goat breeds serve different purposes. Dairy breeds like the Nigerian Dwarf or Alpine are excellent for milk production, while meat breeds like the Boer are raised for their meat. If you're interested in fiber, consider the Angora or Cashmere breeds.

- **Goat Shelter:** Goats need protection from the elements, but their shelter doesn't need to be overly elaborate. A simple, sturdy barn or shed will do, and goats enjoy having a space to climb or play in. Since they are curious animals, they need a secure fence to prevent them from escaping.

For example, Sarah and Tom live on a small homestead where they raise two Nigerian Dwarf goats. They chose this breed for its manageable size and excellent milk production. Their goats provide just enough milk for their family's needs, and the excess is used to make cheese and soap. They built a simple, well-ventilated goat shelter with an open space where their goats can graze freely during the day. This minimal approach allowed them to enjoy the benefits of goat farming without overwhelming their resources.

3.   Bees: Pollination and Honey

Beekeeping is one of the most rewarding ways to contribute to a homestead's ecosystem while providing valuable resources like honey, beeswax, and propolis. Bees are crucial for pollination, helping plants on your homestead thrive and produce better yields. Raising bees also allows you to harvest honey, a delicious and natural sweetener that can be used in cooking, medicine, and skincare.

Considerations for Keeping Bees:

- **How Many Hives?** Starting with one or two hives is typically enough for a beginner beekeeper. You don't need to keep many hives to reap the benefits of pollination or honey production. A single hive can produce several gallons of honey per year.

- **Beekeeping Equipment:** Beekeeping requires some basic tools, including a hive, smoker, and protective gear. However, the minimalist approach to beekeeping is to keep equipment simple and efficient, investing only in what you truly need.

- **Maintenance:** While beekeeping can be rewarding, it also requires regular attention. Bees need to be monitored for diseases, pests, and the availability of food sources. However, once the hive is established, the amount of work required for upkeep is minimal.

For example, the Harris family, living on a suburban homestead, decided to start beekeeping as part of their minimalist homesteading journey. They started with one hive, using a simple Langstroth hive design. Over time, they learned the basics of beekeeping and now harvest honey twice a year. The bees help pollinate their garden, providing them with an abundance of fruits and vegetables. By keeping their beekeeping setup simple and efficient, the Harris family enjoys honey while contributing to the health of their ecosystem.

4.   Beyond Chickens, Goats, and Bees

While chickens, goats, and bees are some of the most common animals kept on homesteads, there are many other animals that can serve specific purposes on the minimalist homestead. Rabbits, for example, are an excellent source of protein and require very little space. Ducks can provide eggs and help with pest control. Even a small herd of pigs can contribute to waste reduction by consuming food scraps and providing meat.

The key is to keep the number of animals in proportion to your needs. Only raise animals that will genuinely contribute to your homestead, and ensure that you have the resources and space to care for them properly. By choosing animals that align with your homesteading goals and maintaining a small, manageable number, you can avoid the burden of over-extending your homestead.

Ethical Animal Care with Minimalist Inputs

Raising animals on a homestead requires an ethical approach that ensures the well-being of the animals while minimizing the resources and inputs required to care for them. The minimalist mindset encourages homesteaders to keep animal care simple, focusing on the basics and avoiding unnecessary expenditures. It also emphasizes respect for the animals' natural behaviors and needs, ensuring that their care is sustainable and humane.

1.   Simple, Sustainable Practices

Caring for animals in a minimalist way involves focusing on their most essential needs: food, shelter, and health. Providing high-quality feed, a safe and comfortable shelter, and proper veterinary care doesn't have to involve expensive or complicated systems. Often, the best care involves creating an environment that allows the animals to thrive naturally.

For example, goats are known to enjoy grazing, and many homesteaders allow their goats to roam freely in a pasture or garden, which reduces the need for purchased feed. Chickens are often free-range, foraging for insects and plants, which provides them with a more natural diet. This approach not only saves money but also encourages the animals to express natural behaviors.

2. Veterinary Care and Health

While minimalist homesteading emphasizes low-input care, it's important not to sacrifice the health and well-being of the animals. Regular health checks, vaccinations, and pest control (such as deworming for goats) are essential to ensure that animals stay healthy. However, the minimalist approach to veterinary care encourages homesteaders to rely on preventive measures rather than expensive treatments.

For instance, when raising bees, it's important to monitor the hive for common pests like varroa mites, which can harm the colony. By regularly inspecting the hive and maintaining a clean environment, beekeepers can reduce the need for chemical treatments, ensuring that the honey and beeswax they harvest are organic and safe.

Feed, Shelter, and Waste in Small-Scale Systems

Providing appropriate feed, shelter, and waste management systems are key aspects of raising animals in a minimalist and sustainable way. By simplifying these processes, homesteaders can ensure that animals are cared for while minimizing resource consumption.

1. Feed

Providing feed that aligns with your animals' natural diets is crucial for maintaining health. For chickens, this might mean supplementing their foraged food with a small amount of grains or pellets. Goats might graze on pasture and be supplemented with hay in the winter. Bees rely on nectar from flowers, but in times of scarcity, homesteaders can feed them sugar syrup or honey.

By growing supplemental feed (such as growing fodder for goats or planting herbs for chickens), homesteaders can reduce the need for store-bought feed, which can be expensive and resource-intensive.

2. Shelter

Animal shelter should be simple yet effective, providing protection from the elements and predators. For example, chickens need a dry, safe place to sleep at night, but a minimalist chicken coop doesn't need to be elaborate. A small, sturdy coop with space for the hens to roost and nest is sufficient.

Goats need a shelter to protect them from extreme weather, but again, a minimalist goat barn doesn't need to be large. A simple three-sided structure with bedding for warmth is often enough to keep them comfortable.

3.   Waste Management

Waste management on the homestead is a key element of ethical and sustainable animal care. For most animals, waste can be composted and used as fertilizer. Chicken manure, for example, is rich in nitrogen and can be composted to enrich soil. Goat manure is also an excellent addition to compost piles.

Composting waste not only reduces the environmental impact of animal husbandry but also helps recycle nutrients back into the soil, completing the natural cycle.

Conclusion

Animals play an essential role in minimalist homesteading, contributing to the sustainability and self-sufficiency of the homestead. By choosing animals that align with your needs, raising them ethically, and providing simple, functional care, you can create a balanced and low-maintenance system that works in harmony with nature. Chickens, goats, bees, and other animals can provide valuable resources while minimizing the strain on your time, space, and resources. By adopting a minimalist approach to animal care, you ensure that your homestead remains sustainable, efficient, and deeply rewarding, allowing you to live a simple and intentional life.

CHAPTER 8

The Minimal Kitchen – Simple Food, Deep Nourishment

In the minimalist homesteading lifestyle, the kitchen becomes more than just a place to cook—it's the heart of the home. It is where nourishing food is prepared, simple meals are shared, and the foundation of healthy living is laid. The minimalist kitchen revolves around a few key principles: simplicity, sustainability, and nourishment. This chapter will explore how to cook from scratch with ease, maintain a minimalist pantry with essential staples, preserve food to make the most of harvests, and plan meals in a way that saves time and money while ensuring deep nourishment.

Cooking from Scratch, the Easy Way

Cooking from scratch is a cornerstone of the minimalist kitchen. It's a way of preparing food that emphasizes whole ingredients and simple methods, cutting out processed foods that are often full of unnecessary additives and unhealthy fats. Contrary to the myth that cooking from scratch is time-consuming and difficult, it can be incredibly simple, cost-effective, and deeply satisfying.

1.   The Benefits of Cooking from Scratch

Cooking from scratch has numerous benefits. First and foremost, it gives you complete control over the ingredients that go into your food. You can choose organic, local, or sustainably sourced ingredients, ensuring that the meals you prepare are healthy and nourishing. Additionally, cooking from scratch often results in less waste, as you can buy ingredients in bulk and use them efficiently, minimizing the need for pre-packaged foods and reducing packaging waste.

By cooking from scratch, you can also save money. Convenience foods like frozen dinners, pre-packaged sauces, or canned soups may seem inexpensive, but over time, they add up. Preparing meals with raw ingredients is often far cheaper, especially when you can source staples in bulk or from local producers.

2. Simple Cooking Techniques

Cooking from scratch doesn't need to involve complex or time-consuming recipes. The minimalist kitchen embraces the philosophy of making meals that are nourishing, tasty, and quick to prepare. The key is to focus on a few versatile recipes that can be easily adapted with different ingredients.

For example, simple one-pot meals like soups, stews, and casseroles can be made with minimal ingredients. Start with a base of broth or stock, and then add seasonal vegetables, grains, or legumes. Add your protein of choice, whether it's chicken, beans, or lentils, and let it simmer to develop deep flavors. These meals are not only easy to prepare but also affordable, and leftovers can be used for additional meals throughout the week.

Stir-fries, salads, and grain bowls are also simple, nutritious meals that require very little prep work. With a few basic ingredients—grains (like rice or quinoa), vegetables, and a protein source—you can whip up a quick and satisfying meal without needing a large variety of ingredients or complex recipes.

For example, Sarah has learned to make a nourishing vegetable soup using whatever seasonal vegetables she has available. She starts with onions, garlic, and a good broth, adding carrots, potatoes, and beans, and then seasoning it with herbs like thyme or rosemary. It's a filling, satisfying meal that she can make in under an hour, using ingredients she likely already has in her pantry.

3.   Batch Cooking: Preparing Simple Meals in Advance

Batch cooking is a minimalist approach to cooking that maximizes efficiency by preparing large quantities of food at once. This method saves time throughout the week, as you can cook once and enjoy multiple meals. It's especially useful for staple dishes like soups, casseroles, and roasted vegetables, which can be made in large batches and stored for later use.

For example, when John and Lisa have a weekend day free, they spend a few hours cooking large batches of their favorite meals, like chili, pasta sauce, and roasted vegetables. These dishes can be stored in the fridge or freezer and eaten throughout the week, reducing the amount of time they need to spend in the kitchen.

Batch cooking doesn't mean spending all day in the kitchen. It's about using a little extra time to prepare foods that can be quickly reheated or assembled into meals. This can make weekday dinners less stressful, as you already have the components ready to go.

Pantry Staples and Batch Prep

A minimalist pantry is one of the most powerful tools in a minimalist kitchen. By focusing on a small selection of versatile, long-lasting staples, you can create a wide variety of meals without overloading your kitchen with unnecessary ingredients. Stocking your pantry with essential staples allows you to make meals quickly and efficiently without the need for complicated shopping lists or excess clutter.

1.   Essential Pantry Staples

A minimalist pantry is built on a few key ingredients that can be used in a variety of dishes. Here's a list of some pantry staples that every minimalist kitchen can benefit from:

- **Grains:** Rice, quinoa, oats, and pasta are versatile staples that serve as the base for many meals. They are long-lasting, easy to cook, and can be used in soups, casseroles, stir-fries, or baked dishes.

- **Legumes:** Beans, lentils, and chickpeas are excellent sources of protein and fiber. They can be used in soups, stews, salads, or as a meat substitute in various dishes.

- **Canned Goods:** A few well-chosen canned goods, such as tomatoes, coconut milk, and beans, can serve as the foundation for a variety of dishes. Choose high-quality, BPA-free cans, and stock up when these items are on sale.

- **Nuts and Seeds:** Almonds, walnuts, chia seeds, and flaxseeds are great for snacking, adding to smoothies, or incorporating into baked goods. They also provide healthy fats and protein.

- **Spices and Herbs:** A small collection of dried herbs and spices can elevate any dish. Consider basic spices like salt, pepper, garlic powder, cumin, chili powder, and Italian seasoning. Fresh herbs can be grown in your garden or purchased in small amounts.

- **Oils and Vinegars:** Olive oil, coconut oil, and apple cider vinegar are key ingredients for dressings, cooking, and marinating.

By keeping a minimalist pantry stocked with these staples, you can quickly assemble a variety of meals without feeling overwhelmed by choices or clutter.

2.   Batch Prep for the Pantry

Batch prepping pantry staples involves cooking or preparing ingredients in advance, so you can easily incorporate them into meals throughout the week. For example, cooking a large pot of beans, roasting a big batch of sweet potatoes, or chopping vegetables and storing them in the fridge can make meal prep much faster.

When Mark and Rebecca have extra time, they batch prep several ingredients for the week. They cook a large pot of quinoa, roast a tray of vegetables, and prepare some chicken breasts in advance. These items can then be used to make salads, bowls, or grain-based dishes, cutting down on the time spent preparing meals during the week.

Preserving: Canning, Drying, and Fermenting

Preserving food is an essential part of the minimalist kitchen. It allows you to make the most of seasonal produce, reduce waste, and ensure that you always have healthy options available. Whether you're preserving fresh fruits and vegetables from your garden or simply stocking up on food when it's in abundance, methods like canning, drying, and fermenting can help you create a long-lasting food supply.

1.   Canning: Preserving the Harvest

Canning is an age-old method of preserving food that involves sealing food in jars and heating it to kill bacteria, allowing it to be stored safely for months or even years. Canning is ideal for preserving fruits, vegetables, jams, and even meats. It is a great way to ensure that the produce from your garden doesn't go to waste and that you have homemade, nutritious options available during the off-season.

For instance, Linda and Jim, who have a large garden, can their tomatoes each year to make homemade marinara sauce. They also can peaches and apples, which they use in baked goods or as a topping for oatmeal. By canning their harvest, they can enjoy the fruits of their labor throughout the year, without worrying about spoilage.

2.   Drying: A Simple and Space-Saving Method

Drying is another excellent way to preserve food. Dried fruits, vegetables, and herbs take up far less space than canned goods and can last for months without refrigeration. Dehydrators are an efficient way to dry produce, but you can also dry food naturally in the sun or using your oven at a low temperature.

For example, Chris uses his dehydrator to dry tomatoes from his garden, which he later uses in soups and sauces. He also dries herbs like basil, thyme, and oregano, which he uses for seasoning during the winter months when fresh herbs are unavailable.

3.   Fermenting: A Gut-Healthy Option

Fermentation is not only a way to preserve food, but it also provides beneficial probiotics that support gut health. Fermented foods like sauerkraut, kimchi, and pickles are easy to make and can be added to meals to enhance flavor and nutrition.

For instance, Anna has embraced fermenting as part of her minimalist kitchen approach. She makes her own sauerkraut using cabbage, salt, and water. Fermented foods like sauerkraut can be stored in jars for several months, and they provide a tangy addition to salads, sandwiches, and meals.

Meal Planning for Sanity and Savings

Meal planning is an essential practice for maintaining a minimalist kitchen and reducing food waste. By planning meals ahead of time, you can ensure that you're using up ingredients efficiently, saving money, and reducing the stress of figuring out what to make at the last minute.

1. Meal Planning Basics

Meal planning doesn't have to be complicated. The key is to create a flexible plan that fits your schedule and allows for variety. Start by planning meals that use similar ingredients, so you don't have to purchase a large number of different items. For example, you could plan a week's worth of meals that all incorporate beans, rice, and vegetables. This not only saves time but also reduces food waste.

2. Batch Cooking for Meal Planning

Batch cooking can be a game-changer when it comes to meal planning. By preparing a large batch of food on one day, you can enjoy multiple meals throughout the week without spending extra time in the kitchen. For example, cook a large pot of soup or stew, then portion it out into containers for later use. Similarly, roast a large tray of vegetables or cook a large batch of grains to use throughout the week.

Conclusion

The minimalist kitchen is all about simplicity, sustainability, and nourishment. By focusing on cooking from scratch, maintaining a minimalist pantry, preserving food, and planning meals efficiently, you can create a kitchen that supports your health and homesteading goals. The minimalist approach to cooking doesn't mean sacrificing flavor or variety—it's about focusing on what you truly need, making the most of your resources, and nourishing yourself and your family with wholesome, simple food. By embracing these principles, you can enjoy a simpler, more intentional kitchen that brings joy and deep nourishment to your life.

CHAPTER 9

Making and Mending – DIY the Minimalist Way

One of the most empowering aspects of minimalist homesteading is the ability to make and mend things yourself. In a world that often encourages consumerism and the constant pursuit of new products, the minimalist approach offers a refreshing alternative: DIY. By embracing simple, practical skills, homesteaders can save money, reduce waste, and create or repair the things they need without relying on external sources. This chapter will guide you through the tools every minimalist homesteader truly needs, basic DIY skills that are incredibly valuable, and how to discern when to do it yourself and when to simplify or seek external help.

Tools Every Minimalist Homesteader Actually Needs

As a minimalist homesteader, it's important to have the right tools that can help you build, repair, and maintain your homestead without accumulating an overwhelming collection of equipment. The minimalist mindset teaches us to invest in high-quality, versatile tools that can perform multiple functions, rather than filling our sheds with specialized tools for every possible task.

1.   Basic Hand Tools

When it comes to hand tools, simplicity is key. Focus on essential tools that will allow you to perform a variety of tasks. Some of the most important hand tools include:

- **Hammer:** A sturdy, all-purpose hammer is indispensable for a variety of tasks, from building structures to making small repairs around the homestead.

- **Screwdrivers:** Both flathead and Phillips screwdrivers are necessary for assembling furniture, fixing appliances, or performing other minor repairs.

- **Pliers:** A good pair of pliers can help with gripping, bending, and cutting wire, among other tasks.

- **Utility Knife:** A sharp utility knife is essential for cutting and trimming various materials such as plastic, wood, and fabric.

- **Tape Measure and Level:** Accurate measurements are crucial for any DIY project, so having a tape measure and level ensures that everything you build or install is straight and properly sized.

- **Wrenches:** A set of adjustable wrenches is invaluable for tightening bolts and nuts on plumbing fixtures, appliances, and other items around the homestead.

- **Handsaw or Circular Saw:** For cutting wood and other materials, a handsaw or a small circular saw can handle most DIY projects, from building furniture to small construction tasks.

- **Shovel and Rake:** Whether for gardening, digging, or maintaining your homestead, a good shovel and rake are indispensable for moving soil, compost, or clearing debris.

2.   Power Tools for Bigger Jobs

While hand tools are excellent for smaller projects, power tools can save time and energy for larger, more demanding tasks. A minimalist homesteader should invest in versatile power tools that can perform several tasks, rather than purchasing a separate tool for every function.

- **Cordless Drill:** A high-quality, cordless drill is essential for driving screws and drilling holes in wood, metal, and plastic. A drill with multiple bits will serve you for almost any task, from assembling furniture to building structures.

- **Circular Saw:** A circular saw is perfect for cutting wood for construction projects, including building raised garden beds, fencing, or furniture.

- **Jigsaw:** A jigsaw is great for making curved cuts or intricate patterns in wood or other materials. It's a highly versatile tool that can be used for a variety of DIY tasks.

- **Sander:** For smoothing rough wood surfaces, a sander can help you finish woodworking projects without having to use manual methods like hand-sanding, which can be tedious and time-consuming.

- **Generator:** If you live off-grid, a small portable generator can be invaluable for providing backup power when solar or wind energy isn't enough to meet your needs.

3.   Basic Maintenance and Repair Tools

No matter how carefully you design your homestead or how durable your equipment is, things will break or need maintenance over time. Minimalists seek to repair and maintain what they already own, rather than constantly replacing items. A set of basic repair tools will help you address common problems.

- **Needle and Thread:** For mending clothes, tents, and other fabric items, having a needle and thread is essential. A simple sewing kit with various needles, threads, and fabric patches can help you fix tears, add buttons, or make basic alterations to clothing.

- **Duct Tape and Superglue:** Duct tape is one of the most versatile repair tools available. It can be used to patch up leaks, hold things in place, and temporarily fix broken items. Superglue is also essential for making small, quick repairs to items that need bonding.

Investing in high-quality, durable tools will serve you well as a minimalist homesteader. Choose tools that are built to last, and focus on those that are most versatile, so you aren't burdened with a collection of items you rarely use.

Simple Skills: Sewing, Sharpening, Fixing

The minimalist mindset is about embracing simplicity, not just in your environment but in your daily activities. One of the best ways to live simply is to learn basic skills that allow you to make, mend, and fix things on your own. By developing practical skills like sewing, sharpening, and basic repairs, you can reduce your dependence on outside services and foster a sense of independence and satisfaction.

1.   Sewing: Repair and Reuse

Sewing is an essential skill for any minimalist homesteader. Whether you need to repair a torn shirt, alter a pair of pants, or create your own clothing from fabric, knowing how to sew can save you money and reduce waste.

- **Hand Sewing:** Even if you don't own a sewing machine, hand sewing is a valuable skill that allows you to make simple repairs. Learning how to sew on buttons, mend seams, and patch small holes can extend the life of your clothes, reducing the need to buy new ones.

- **Machine Sewing:** For larger projects, such as making curtains, quilts, or even new clothing, a sewing machine can be invaluable. A basic, manual machine is often sufficient for most homesteading tasks, and it can be used to create custom items, from bags to bedding, to fit your specific needs.

For example, Samantha, living on her minimalist homestead, learned to sew to mend clothes and create custom items for her family. She no longer had to purchase new items when things tore, and she was able to make beautiful homemade gifts and functional items like pillowcases, aprons, and reusable produce bags.

2.   Sharpening: Tools and Knives

Maintaining sharp tools is one of the best ways to extend their lifespan and make your work easier. Whether you're sharpening knives, garden tools, or other implements, knowing how to maintain your tools with minimal effort will help you avoid buying replacements. A sharp tool is also safer to use, as it reduces the likelihood of accidents.

- **Knife Sharpening:** A sharp knife is essential for preparing food, working with wood, and other homesteading tasks. You can use a sharpening stone, honing rod, or an electric sharpener to keep your knives in good condition.

- **Tool Sharpening:** Shovels, hoes, and garden shears require regular sharpening to keep them effective and make tasks like weeding or digging easier. A simple sharpening stone or file can be used to keep these tools in top condition.

By learning to sharpen your own tools, you can reduce the need for professional sharpening services and extend the life of your essential equipment.

3.   Basic Fixing: Household Repairs

Fixing small problems around the house is a key skill for any minimalist. From mending furniture to fixing broken pipes, learning basic repair skills can save money and reduce waste. Minimalists seek to repair and repurpose rather than throwing things away or buying new items.

- **Furniture Repair:** Wooden furniture can often be repaired with basic tools like a hammer, nails, wood glue, and clamps. Knowing how to fix wobbly chairs, cracked tables, or loose drawers can prevent you from needing to buy   new   furniture.

- **Plumbing and Electrical:** While complex plumbing or electrical issues should be left to professionals, minor repairs can often be tackled by a homeowner. Learning how to fix leaky faucets, replace washers, or install light fixtures can save you money on service calls.

- **General Maintenance:** Regular maintenance is the key to preventing bigger problems down the road. Whether it's fixing a leaking roof or repainting a fence, addressing small issues before they grow into major repairs is an essential skill for a minimalist homesteader.

For example, Ethan, a self-sufficient homesteader, took pride in fixing his own appliances. When his washing machine broke, he didn't call a repairman. Instead, he researched the issue, ordered a new part, and fixed the machine himself. This approach saved him money and prevented the waste of throwing out the entire appliance.

When to DIY and When to Simplify

While DIY is often empowering and satisfying, minimalist homesteaders understand that it's not always the most practical or efficient approach. The key to minimalist DIY is knowing when to take matters into your own hands and when to simplify by seeking help or outsourcing certain tasks.

1. When to DIY

DIY is ideal when it saves you money, gives you a sense of accomplishment, or allows you to repurpose materials that would otherwise go to waste. If a task is simple and you already have the tools and skills, then doing it yourself is a great option.

For instance, if a small piece of furniture breaks or a shelf falls off the wall, it's likely easier and cheaper to fix it yourself than to hire a professional. Similarly, if you need to sew a patch on a pair of jeans or make a few adjustments to your wardrobe, these tasks can be done in a few minutes with basic sewing tools.

2. When to Simplify

However, not every task is worth tackling yourself. If a project is time-consuming, requires specialized knowledge, or involves safety risks, it's often more efficient and safer to simplify by hiring a professional. For example, complex electrical or plumbing issues should be handled by experts to avoid potential hazards or costly mistakes. Similarly, if a project will take up too much of your time, it might be worth simplifying by outsourcing it.

For example, while DIY repairs to your fence or garden shed are likely manageable, hiring a professional to install a new septic system or handle major structural repairs may be a better option for long-term success and safety.

Conclusion

The minimalist approach to DIY encourages independence, self-reliance, and practicality. By learning simple skills such as sewing, sharpening, and basic repairs, minimalist homesteaders can save money, reduce waste, and create a home that aligns with their values. At the same time, minimalist DIY is about understanding when to take on tasks yourself and when to simplify by seeking professional help. With the right tools, skills, and mindset, DIY can be a deeply rewarding part of your minimalist homesteading journey, fostering creativity, sustainability, and self-sufficiency in every aspect of life.

CHAPTER 10

Money, Bartering, and the Simple Economy

Minimalist homesteading is often about finding ways to live sustainably, without unnecessary complexity or the need for excessive material wealth. However, the reality is that money and resources still play a significant role in sustaining our lifestyles and making homesteading a viable option. The minimalist homesteader seeks a balanced relationship with money—one that doesn't dictate every decision, but rather serves as a tool for maintaining independence and freedom. In this chapter, we will explore how to earn from the homestead without burning out, the power of bartering and community sharing, and how to budget in a minimalist way for a fulfilling rural life.

Earning from the Homestead Without Burning Out

One of the key aspects of minimalist homesteading is finding ways to generate income from your land or homestead, but in a manner that aligns with the values of sustainability and simplicity. It's all too easy to get caught up in the need to hustle for money, working long hours to sell produce, products, or services, only to find that you've created a new source of stress rather than a fulfilling, peaceful life. The minimalist homesteader seeks income generation that is sustainable—both in terms of resources and personal well-being.

1.   The Balance Between Work and Life

For many homesteaders, the goal is not to maximize income, but to create enough revenue to cover basic needs and support the lifestyle they desire. This often means living frugally, reducing reliance on outside income, and finding creative ways to generate money that doesn't detract from the overall purpose of the homestead: simplicity.

Ways to Earn Income on the Homestead:

- **Growing and Selling Produce:** If you have the space, selling excess fruits, vegetables, and herbs can be a simple and profitable way to earn money. Small-scale farmers or gardeners often sell directly to consumers at farmers' markets or through Community Supported Agriculture (CSA) programs.

- **Livestock and Animal Products:** Chickens can provide eggs, goats can offer milk, and bees can produce honey. Selling these products, or even starting a small dairy or honey business, can generate steady income without requiring too much extra effort.

- **Crafts and Handmade Goods:** Many homesteaders use their skills to create handmade goods, whether it's soap, candles, wool products, or woodworking items. Selling these products locally or online through platforms like Etsy can provide a small but steady income.

For example, the Johnson family lives on a small homestead where they grow a variety of vegetables and sell surplus produce at their local farmer's market. Instead of farming full-time, they focus on cultivating just enough to generate supplemental income, which allows them to maintain a relaxed pace of life. This balance of earning income and keeping the work manageable is a key part of their minimalist homesteading approach.

2.   Leveraging Skills for Side Hustles

Minimalist homesteading doesn't always have to rely on farming or animal husbandry for income generation. Homesteaders often bring a variety of skills with them, from woodworking to gardening to web design. Leveraging these skills in small, side hustles can provide additional income without overwhelming your homestead's rhythms.

For example, Rachel is an experienced carpenter who lives on her homestead with her family. She spends a few hours each week building custom furniture or small structures like garden boxes, which she sells locally. She also offers carpentry workshops for people in the community interested in learning the craft. These small side jobs provide supplemental income but do not overwhelm her time or energy. By integrating her skills into the homestead's routine, she avoids the burnout that often comes with taking on too much.

Barter, Trade, and Community Sharing

One of the core principles of minimalist living is reducing reliance on external systems, including financial ones. While money is still necessary for some purchases, the practice of bartering, trading, and community sharing offers an alternative to the traditional economy. By embracing these methods, homesteaders can access goods and services without having to rely solely on cash transactions, creating stronger bonds with neighbors and building a more resilient local economy.

1.   Bartering and Trading

Bartering, the act of exchanging goods or services without money, has been a central part of human economies for centuries. On the homestead, bartering can be a valuable way to obtain the items or services you need without spending money. It is a practical approach that emphasizes self-reliance and community connection, and it encourages homesteaders to think creatively about what they can offer in exchange.

Examples of Bartering on the Homestead:

- **Farming or Livestock Products:** If you raise animals, you might barter eggs, milk, or meat with your neighbors in exchange for something they have—such as firewood, tools, or assistance with building projects.

- **Skills for Goods or Services:** If you have a skill like carpentry, gardening, or sewing, you can offer those services in exchange for items or help with tasks that you don't know how to do. For example, a neighbor who knows how to make cheese might offer to trade cheese for some freshly harvested vegetables from your garden.

● **Shared Labor:** Sharing labor is a valuable form of barter. If you're working on a major project, such as building a shed or repairing a fence, you might offer your skills or time in exchange for help with the work. This fosters a strong sense of community and helps you maintain your homestead without overextending yourself.

For example, Tim and Claire have a small farm and barter with neighbors for goods and services. They often trade vegetables for honey, firewood for dairy products, or even their own labor in exchange for help with various building projects. This system allows them to meet their needs without spending money and strengthens their connection with the local community.

2.   Community Sharing and Co-ops

Sharing resources within a community creates a safety net that benefits everyone involved. In rural or homesteading communities, shared tools, land, or skills can significantly reduce individual costs and increase collective resilience. Community sharing could include things like shared garden plots, tool libraries, or local food cooperatives (co-ops).

For instance, a group of homesteaders may create a cooperative garden where each person contributes labor or materials, and in return, they share the harvest. Similarly, a local tool library may allow homesteaders to borrow expensive equipment that they don't need on a regular basis, such as tractors or sawmills, without having to invest in purchasing it themselves.

Community sharing is a powerful way to reduce consumption, foster relationships, and make homesteading more affordable for everyone. It aligns with the minimalist principle of avoiding excess and living with what you truly need, while still maintaining a high quality of life.

3.   Time Banking

Time banking is a community-based system where individuals exchange hours of service rather than money. In a time bank, one hour of work (such as babysitting, carpentry, or dog walking) is exchanged for one hour of another service (like gardening or transportation). This system can be incredibly effective for homesteaders who have a variety of skills to offer and who want to help others while receiving help in return.

For example, the Smiths, living in a rural area, participate in a time bank where they exchange a few hours of gardening for help with their animal care. This system is particularly helpful during busy seasons, such as spring planting or harvest time, when homesteaders may need extra hands to manage their workload.

Minimalist Budgeting for Rural Life

A minimalist homesteader's budget should be aligned with the values of simplicity, sustainability, and self-reliance. Minimalist budgeting doesn't mean living without; it's about being intentional with how you spend your money and eliminating unnecessary expenses. In rural homesteading, where many expenses are related to land maintenance, equipment, and self-sufficiency, budgeting is an essential tool to ensure long-term financial sustainability.

1.   Understanding Your Expenses

Before creating a budget, it's important to understand where your money goes. A minimalist approach to budgeting starts with identifying essential and non-essential expenses. Essentials may include:

- **Land Maintenance:** Costs associated with maintaining the property, such as tools, seeds, fertilizers, and livestock care.

- **Utilities:** While many homesteaders rely on alternative energy sources (like solar power or wind), traditional utilities (electricity, water, etc.) still play a role in many homesteads.

- **Health and Insurance:** Medical expenses and insurance coverage are essential to protect your health and assets.

Once you've accounted for essentials, the goal is to eliminate non-essential expenses that don't contribute to your core values. For example, rather than spending money on dining out or buying trendy clothes, a minimalist homesteader might focus on purchasing quality tools or investing in long-term sustainability projects on the homestead.

2.   Sustainable, Long-Term Investments

Minimalist budgeting for rural life also involves making sustainable investments that will pay off over time. For example, investing in high-quality tools that last longer or planting perennial crops (which grow year after year) can save money in the long run. Similarly, making improvements to the homestead, such as building an energy-efficient home or installing rainwater harvesting systems, may require an upfront investment but will lead to reduced costs over time.

For example, Emily and Peter, living on their homestead, invested in a high-quality wood stove for heating their home. While the initial cost was significant, they now save money on heating bills every winter, and the stove will last for decades.

3.   Tracking and Planning for Seasonal Variations

Rural life often has seasonal variations in expenses. For example, in the summer, you might need to spend more on irrigation, seeds, and equipment maintenance, while in the winter, fuel and heating costs may rise. By understanding these seasonal fluctuations, you can plan ahead and set aside savings during less-expensive months to cover costs during peak seasons.

Conclusion

The minimalist homesteader approaches money, bartering, and budgeting with intentionality, balancing self-sufficiency with the practical needs of modern life. Earning from the homestead without burning out, engaging in bartering and community sharing, and adopting minimalist budgeting principles ensure that you are living sustainably without financial stress. By understanding the value of time, skills, and resources, you can cultivate a homestead that is both economically sustainable and deeply aligned with your values of simplicity and self-reliance. This approach fosters financial freedom and builds a strong, supportive community, allowing you to thrive within your means and live a life that is rich in purpose and contentment.

CHAPTER 11

Rhythms, Routines, and Seasons

L iving a minimalist homesteading lifestyle means embracing the rhythms of nature—understanding the flow of seasons and aligning your daily routines to them. This chapter explores how connecting with the natural world's cycles can help you live more intentionally, structure your day with simplicity, and enjoy the calm beauty of slower mornings and earlier nights. By adopting a rhythm that is in harmony with the seasons, your work, energy, and peace of mind can be more balanced and sustainable, allowing you to cultivate a life of contentment and purpose.

Embracing Seasonal Living

One of the key aspects of minimalist homesteading is learning to live with the seasons. This means adjusting your activities, routines, and mindset according to the changing weather patterns and the natural cycles of the earth. Seasonal living involves honoring the rhythms of the year, understanding what is required in each season, and simplifying your tasks accordingly. It's a way of living that focuses on being present and mindful of the environment, rather than fighting against it or trying to override its natural pace.

1.   Spring: The Season of Renewal and Growth

Spring is a time of renewal and planting. It is the season when you begin to plant the seeds of the year, both literally and figuratively. In the garden, this is the time to prepare your soil, plant your early crops, and start your seeds indoors if necessary. It's a time to focus on new beginnings, fresh energy, and setting the foundation for the year ahead.

For example, the Richardson family begins each spring by prepping their raised garden beds, ensuring the soil is rich and fertile. They start planting cool-weather crops like peas, lettuce, and spinach, and spend time outdoors preparing for the more intense labor of summer. This approach not only ensures a successful growing season but also reflects their minimalist lifestyle—each action in spring is purposeful and aligned with the seasons, without overcomplicating the process.

Spring is also a time for mental and emotional renewal. It's a time to set intentions for the year, embrace new projects, and begin planning for the rest of the seasons ahead. This aligns perfectly with minimalist homesteading because it encourages you to act intentionally, setting goals that resonate with your deeper values.

2.   Summer: The Season of Abundance and Harvest

Summer is often the most active season on a homestead. With longer days and warmer weather, there is plenty of time for growing, tending, and harvesting. This is the season when your efforts in spring begin to bear fruit, and the homesteader reaps the rewards of their labor. The minimalist approach to summer living encourages you to focus on what's truly necessary and avoid overcommitting to tasks or projects that drain your energy.

During the summer, homesteaders focus on tending their gardens, harvesting crops, preserving food, and keeping up with the day-to-day maintenance of the homestead. For instance, Mary and Tom have a small garden where they grow tomatoes, beans, and peppers. Throughout the summer, they harvest and preserve what they can—canning tomatoes, drying herbs, and freezing excess produce. They avoid the temptation to overplant and focus on crops that will truly nourish their family. By doing so, they enjoy the fruits of their labor without overwhelming themselves with unnecessary tasks.

Summer is also a time for community engagement. Many homesteaders participate in farmers' markets or share their excess produce with neighbors. This reflects the minimalist value of sharing abundance without hoarding, contributing to a local economy based on cooperation and generosity.

3.   Fall: The Season of Reflection and Preparation

Fall is a transitional season, where the abundance of summer begins to wane, and the homesteader turns inward, both physically and mentally. It's the time to prepare for the winter months, harvest late-season crops, and reflect on the work done throughout the year. Minimalists use this time to organize their homes, store food, and begin making plans for the colder months.

For example, David and Sarah spend their fall months preparing for winter. They harvest their root vegetables, dry herbs, and preserve what is left of their harvest in jars. They also take time to clean and organize their homestead, storing tools, firewood, and blankets for the colder months. Fall is also a time of slowing down—without the intensity of summer's demands, the minimalist homesteader has time for reflection, reading, and introspection.

In terms of mindset, fall is about letting go. It's a season for paring down—not just physically, but emotionally and mentally as well. It's a time to shed excess and simplify further, making space for what truly matters. For many, fall becomes a time for letting go of old projects, focusing on what worked well, and simplifying life even more.

4.   Winter: The Season of Rest and Renewal

Winter can be the most challenging season for a minimalist homesteader, but it is also the time for rest, renewal, and simplicity. With fewer outdoor tasks and colder weather, winter is the season to slow down, rest, and recharge for the coming year. This period of rest is essential to avoid burnout and maintain balance in the cycle of life.

Winter provides time to reflect, focus on indoor projects, and enjoy time with family. For example, during winter, Rebecca and her family take advantage of the downtime to learn new skills like knitting, quilting, or woodworking. They spend time planning their garden for the next season, making sure that they are prepared for spring. They also focus on indoor projects like organizing, deep cleaning, and maintaining the tools and equipment used during the rest of the year.

Winter is also a season to reduce external distractions and focus inward. Minimalist homesteaders often use this time to simplify further—whether by decluttering, reducing commitments, or focusing on personal growth and development.

Structuring Your Day the Low-Tech Way

In the modern world, technology can be overwhelming. It often leads to distractions, stress, and a sense of time slipping away. Minimalist homesteading, however, calls for embracing a slower, more intentional pace of life. This means structuring your day in a way that focuses on important tasks, reduces stress, and allows time for rest and relaxation, all without relying heavily on technology.

1.   The Power of a Simple Daily Routine

One of the most powerful ways to bring simplicity into your life is by creating a daily routine that aligns with your goals and values. A minimalist routine doesn't need to be rigid or over-complicated—it's simply about setting a structure that supports productivity, health, and well-being.

For example, Mary and her family begin each day with a morning routine that revolves around chores, outdoor work, and family time. They wake up early to tend to animals, check on the garden, and prepare breakfast. The day then flows with minimal distractions—chores, gardening, and a bit of time for creative projects or hobbies. By structuring their day this way, they maintain a sense of peace and focus throughout the day.

A minimalist routine also helps you prioritize rest. While homesteading requires hard work, it's essential to make time for breaks. By scheduling time for short naps, meals, and relaxation, minimalist homesteaders ensure they have the energy to tackle important tasks while preventing burnout.

2.   Low-Tech Tools and Calendars

Minimalists often shy away from digital tools, opting instead for simple, low-tech systems to organize their days. A paper planner, wall calendar, or a simple to-do list can serve as effective alternatives to digital devices that often lead to distractions.

For instance, Jack and Lily, living on their homestead, use a wall calendar to track important dates—planting schedules, harvest days, and community events. Their paper calendar serves as a central point for family coordination, reducing the reliance on phones or computers to manage their time. This simple system keeps their homestead running smoothly without the distractions that come from constant digital notifications.

Additionally, homesteaders often use traditional methods of time tracking, such as keeping a weather journal or using a sun dial, which connect them more directly to the natural rhythms of the earth.

The Beauty of Slow Mornings and Early Nights

Nicci Brochard & Dr.Ben Chuba

A key element of minimalist living is embracing a slower pace of life. One of the best ways to do this is by starting your day with a peaceful morning routine and winding down with an early night. These rhythms help you stay grounded, reduce stress, and maintain a balanced, fulfilling lifestyle.

1.   Slow Mornings: A Time for Reflection and Intentions

Slow mornings allow you to start the day with clarity and intention. Rather than rushing into the day with a list of tasks, a minimalist morning routine gives you time to center yourself, plan your day, and focus on the things that matter most.

For example, Emma and Paul start their mornings with a quiet cup of coffee on the porch, followed by a short walk to check on their chickens and garden. They take time to reflect on their goals for the day and discuss what needs to be done. This slow morning pace sets the tone for the rest of the day—focused, intentional, and calm.

Slow mornings also provide the opportunity for self-care. Whether it's practicing yoga, reading, or simply enjoying a quiet breakfast with family, taking the time to savor the start of your day can reduce stress and set a positive tone for everything that follows.

2.   Early Nights: Rest and Restoration

Just as slow mornings set the tone for the day, early nights offer a sense of calm and restoration. In minimalist living, it's essential to prioritize rest. By going to bed early, you ensure that your body and mind have the time they need to recharge, preparing you for the demands of the next day.

On the homestead, early nights can be especially important during busy seasons. After a full day of hard work, resting early allows you to wake up refreshed and ready to tackle another day. The darkness of the evening also encourages a natural wind-down, where distractions are minimized, and family time or personal reflection takes center stage.

For instance, Maria and her husband Ben have a routine where they light candles after sunset, enjoy a light meal, and sit down together for a quiet evening of conversation. They find that this slow-paced evening ritual not only promotes relaxation but helps them maintain a deep sense of connection with each other and their homesteading goals.

Conclusion

Living in tune with the rhythms of nature and structuring your day the low-tech way offers a deep sense of peace and fulfillment. Embracing seasonal living, creating simple routines, and allowing for slow mornings and early nights helps minimalist homesteaders avoid burnout, maintain balance, and foster a lifestyle of intentionality and simplicity. These practices allow you to savor the present moment, honor your connection with nature, and live in harmony with the world around you. Ultimately, slowing down and aligning your life with the natural rhythms is one of the most profound ways to create a meaningful, sustainable homesteading lifestyle.

CHAPTER 12

Raising Kids (or Pets!) the Minimalist Homestead Way

Minimalist homesteading is not just about simplifying your own life—it's about fostering an environment where everyone in the household, whether human or animal, can thrive. Raising children or pets on a homestead offers the opportunity to nurture responsibility, independence, and a deep connection to nature, all while cultivating a lifestyle that values simplicity, sustainability, and intentional living. Whether you have young children or furry companions, creating a minimalist homestead that incorporates their needs and well-being into your daily routines and long-term goals can help everyone live in harmony with the rhythms of nature. This chapter explores how to raise children in a way that encourages simplicity, freedom, and connection with the world around them, as well as how to include pets in your off-grid, sustainable life.

Simple Childhood: Nature, Chores, and Freedom

One of the most rewarding aspects of raising children on a homestead is the opportunity to introduce them to a life that values simplicity, connection to nature, and personal responsibility. A minimalist homesteading lifestyle encourages kids to be engaged in the natural world, cultivate a strong work ethic, and learn essential life skills.

1.   Nature-Based Play and Exploration

Children raised on a homestead have the privilege of growing up surrounded by nature, which is an invaluable opportunity for learning and growth. Rather than being confined to indoor activities or glued to screens, kids on a homestead can explore the outdoors, observe wildlife, and play in open spaces. Nature offers endless opportunities for curiosity and creativity, and it fosters a deep sense of wonder and respect for the environment.

For example, the Williams family lives on a 5-acre homestead and believes that outdoor play is one of the most important aspects of their children's upbringing. Their kids, aged 7 and 9, spend hours exploring the woods, playing in the creek, and learning about the different animals on the property. This connection to nature not only provides them with a sense of adventure but also helps them develop a greater appreciation for the earth and its natural rhythms.

Allowing children to roam and explore outdoors, within safe boundaries, builds their self-confidence and independence. It gives them the freedom to discover new things, make mistakes, and learn valuable lessons through hands-on experiences— skills that are hard to replicate in the classroom or through digital screens.

2.   Chores: Building Responsibility and Work Ethic

One of the cornerstones of homesteading is self-sufficiency, and children on the homestead are an integral part of this process. Rather than shielding kids from hard work, minimalist homesteading encourages involving them in chores and tasks that contribute to the household's functioning. Chores provide an invaluable opportunity for children to develop responsibility, teamwork, and a sense of accomplishment.

Children can be involved in age-appropriate chores from an early age. Young children might start with simple tasks, such as feeding the chickens, watering the plants, or gathering firewood. As they grow older, they can take on more complex tasks, such as gardening, harvesting crops, or helping with animal care.

For instance, Anna and Tom involve their two children, aged 5 and 8, in the daily chores on their homestead. Each child has a set of responsibilities, from collecting eggs to helping with the garden. Not only does this teach them the value of hard work, but it also allows them to feel connected to the rhythms of the homestead. By completing tasks that directly contribute to their family's well-being, kids learn how their actions have a positive impact on the household.

Moreover, giving children responsibility in a homesteading setting fosters their sense of independence. They learn how to take care of things, make decisions, and solve problems on their own—all essential life skills that build confidence and maturity.

3. Freedom and Self-Reliance

Minimalist homesteading also allows children the freedom to explore and manage their own time. Rather than being overscheduled with extracurricular activities or being tied to rigid routines, children on a homestead often have the freedom to structure their day around nature's rhythms. This freedom allows them to take ownership of their own learning, whether it's through self-directed play, independent projects, or exploring their own interests.

For example, on the Miller family homestead, their children have a flexible schedule that allows them to spend time learning in the garden, experimenting with new cooking techniques, or building projects in the workshop. The family prioritizes free time for creative activities and outdoor exploration, allowing the kids to follow their curiosity and develop their unique talents. This approach encourages independence, resourcefulness, and a sense of ownership over their time.

Incorporating freedom into their daily lives helps children become more adaptable, resilient, and capable of thinking for themselves, setting them up for a successful future in a world that values adaptability and creative problem-solving.

Homeschooling, Life Skills, and Values-Based Education

One of the most empowering aspects of minimalist homesteading is the ability to educate children in ways that are meaningful, practical, and values-based. Traditional schooling often focuses on academic knowledge, but a minimalist homesteader's approach to education encompasses a more holistic, hands-on model. Homeschooling, life skills development, and values-based education are central to raising children in the minimalist lifestyle, helping them learn how to live sustainably, think critically, and build strong moral foundations.

1.   Homeschooling: Personalized Learning in Tune with Nature

Homeschooling is a natural fit for many homesteading families, as it allows for a more personalized, flexible approach to education. It enables children to learn at their own pace, pursue areas of interest, and incorporate practical, hands-on experiences into their learning process.

For example, the Turner family homeschools their three children on their homestead in the mountains. Rather than following a rigid curriculum, they focus on practical subjects like gardening, animal care, carpentry, and cooking, alongside traditional subjects such as math, reading, and science. The children learn by doing, engaging in real-life projects that integrate academic concepts into daily tasks.

By learning through direct experience, children gain a deeper understanding of how the world works. For instance, when the Turners' children study math, they often apply it in real-world situations—measuring out seed quantities for planting or calculating the area of land to be tilled. Homeschooling in this way helps to build critical thinking skills and fosters a love for learning that is directly connected to their lives.

2.   Life Skills: Preparing for the Future

Minimalist homesteading emphasizes the importance of life skills—practical knowledge that prepares children for adulthood and empowers them to live independently and responsibly. These skills are often learned naturally through participation in everyday tasks and activities around the homestead.

Life skills include things like cooking, sewing, gardening, animal care, woodworking, and basic repairs. Children can learn how to plan meals, manage time, and handle money, which are essential skills for adult life. For example, teaching kids how to preserve food through canning or fermentation provides them with valuable skills they can use for the rest of their lives.

The Evans family, for example, teaches their children basic carpentry skills, such as how to use tools to build simple furniture and structures. Their children also help with tasks like canning tomatoes, collecting eggs, and preparing food. These activities are not only practical but also instill a sense of accomplishment, pride, and independence in the children.

Life skills are more than just practical tasks—they are about fostering self-sufficiency, confidence, and responsibility. By learning these skills from a young age, children become more capable of navigating the challenges of life and building a sustainable future.

3. Values-Based Education: Nurturing Integrity and Compassion

Minimalist homesteading is deeply rooted in values such as sustainability, respect for nature, kindness, and self-reliance.

Raising children in an environment that embodies these values can help instill a strong moral foundation that guides them throughout their lives.

Teaching children about the importance of sustainability, respecting animals, and caring for the land is an essential part of values-based education on the homestead. For example, the Grant family teaches their children to honor the animals they raise and to use every part of the animal for food or other practical purposes. This respect for the animals and the environment is woven into their daily lives and is an important part of their children's education.

Moreover, minimalist homesteading encourages the cultivation of virtues like patience, hard work, and gratitude. Whether it's waiting for crops to grow, repairing broken equipment, or accepting setbacks and challenges, homesteaders learn to embrace the process rather than focusing solely on the outcome. These values, instilled early on, help children develop a strong sense of integrity and character.

Including Furry Companions in Your Off-Grid Plan

Pets can also play an important role in a minimalist homestead, offering companionship, assistance, and even working roles. Whether you have a dog, cat, or even a few farm animals, including your furry companions in the off-grid plan is essential for maintaining a sustainable lifestyle.

1.   Dogs: Protection and Assistance

Dogs are often a key part of the homestead, helping with everything from guarding the property to herding livestock. A well-trained dog can be an invaluable asset, ensuring the security of your home and animals while providing companionship.

For example, the Hart family raises livestock dogs that help keep their sheep and goats safe from predators. Their dogs work alongside them, and the children are actively involved in training and caring for the animals. This partnership teaches kids responsibility, teamwork, and a deep bond with animals.

2.   Cats: Pest Control and Companionship

Cats also have a role to play on the homestead. Not only are they cherished companions, but they also serve as natural pest controllers, keeping rodents and other small animals in check. Many homesteaders find that their cats' natural hunting instincts are helpful for keeping their food stores safe and reducing the need for chemicals or traps.

3.   Other Pets: Providing Love and Connection

While dogs and cats are the most common pets on homesteads, other animals like rabbits, chickens, and even goats can serve as family pets and offer companionship. These animals also teach children valuable lessons in empathy, care, and respect for living creatures.

The Chave family, for instance, has a pet rabbit named Fluff that their children help care for. The children enjoy spending time with Fluff, learning about the animal's needs and developing a sense of nurturing and compassion.

Conclusion

Raising children and pets the minimalist homestead way is about fostering a lifestyle rooted in simplicity, connection, and responsibility. By embracing the natural rhythms of life, teaching life skills, and incorporating furry companions into the homestead, families can create a nurturing, sustainable environment where everyone—both human and animal—can thrive. Whether it's through outdoor play, hands-on education, or the companionship of animals, minimalist homesteading offers the opportunity to raise children and pets in a way that emphasizes freedom, self-reliance, and respect for the world around us. This intentional approach creates a strong foundation for a rich, fulfilling life that values quality over quantity and nature over excess.

CHAPTER 13

Community Without Clutter

Minimalism is often seen as an individual pursuit, focused on simplifying one's own life, but true fulfillment comes when those efforts extend to our communities. Building connections with others and creating strong, supportive networks are integral to living a fulfilling minimalist lifestyle. This chapter will explore how to find and build your local network, host simple gatherings that foster community without overwhelming your space, and harness the power of shared tools, skills, and meals. The minimalist approach to community is about nurturing authentic, meaningful relationships while minimizing excess and clutter, both in terms of physical possessions and unnecessary obligations.

Finding and Building Your Local Network

In minimalist living, the focus is on quality, not quantity, and this philosophy can be applied to your social circles as well. Building a local network is not about collecting a large group of acquaintances but about fostering deep, meaningful connections with those who share similar values, interests, and goals. This network becomes your support system, helping to sustain your homesteading efforts and enriching your life through shared experiences.

1.   Choosing Like-Minded People

Building a minimalist homestead is not just about simplifying your life—it's about creating space for the people who truly matter. For many homesteaders, this means finding others who share a passion for sustainability, self-sufficiency, and simple living. The first step in building a local network is identifying those people in your community or surrounding areas.

Start by attending local farmers' markets, homesteading workshops, or community garden meetings, which are great places to meet like-minded individuals. Many communities also have homesteading groups or local online forums (like Facebook groups or Nextdoor) where people share ideas, resources, and advice. These platforms provide opportunities to connect with others who value sustainability and minimalism.

For example, Sarah and Ben moved to a rural area where they didn't know anyone. They started by attending local farmers' markets and volunteering at a nearby community garden. Over time, they built a small but strong network of like-minded individuals who share similar values around food, sustainability, and the minimalist lifestyle. This network has provided them with invaluable support, whether in the form of advice, bartering goods, or simply sharing ideas for sustainable living.

2.   Collaborating and Sharing Resources

Once you've established a group of individuals who share your interests, collaboration becomes key to sustaining your homestead without overburdening yourself. Many homesteaders rely on community networks for help with tasks that are too large or too specialized for one person to handle. Collaborating allows everyone to share the burden and take advantage of skills and resources that might otherwise be unavailable.

For example, the Harris family, who live on a small farm, have found that by working together with their neighbors, they can share tools, help each other with large tasks, and trade excess produce. They collaborate on building projects, such as constructing a shed, and share equipment like tractors, which would be cost-prohibitive for one family to own. This shared use of resources not only makes life easier but strengthens community ties.

Building a local network also provides an opportunity for support and accountability. Whether you are learning new skills or tackling projects, having a group of people who can offer advice, feedback, or help when needed can make all the difference. For many minimalist homesteaders, community is the foundation that allows them to live intentionally and sustainably without feeling isolated.

Hosting Simple Gatherings

One of the core elements of minimalist living is valuing experiences over things. Hosting gatherings is a wonderful way to build a sense of community while staying true to the minimalist ethos. Simple gatherings focus on connection rather than extravagant decorations or over-the-top planning. They are about sharing time, food, and ideas with others in a relaxed and meaningful way.

1.　Keeping It Simple

When hosting a gathering, whether it's a small dinner party, a potluck, or a community workday, the minimalist approach emphasizes simplicity. Rather than getting caught up in the pressure to create elaborate meals, decorations, or activities, focus on creating a welcoming atmosphere that encourages connection.

For example, the Kelly family frequently hosts casual gatherings where friends and neighbors come together to share a meal. The food is simple—often fresh produce from their garden, baked bread, and homemade soups—but the focus is on good conversation and community. The Kellys avoid overdecorating their home or buying expensive ingredients. Instead, they prioritize making their guests feel at home and ensuring the experience is enjoyable and meaningful.

Similarly, you can host a community potluck where everyone brings a dish to share, reducing the burden on the host while ensuring that the food is varied and plentiful. This encourages sharing and participation, rather than creating a stressful environment where the host feels responsible for every aspect of the gathering.

2.　Celebrating the Seasons

Seasonal gatherings are a wonderful way to align your community celebrations with the rhythms of nature. Whether it's a harvest dinner in the fall, a spring planting celebration, or a summer solstice bonfire, these gatherings honor the cycles of the land and foster a deeper connection to the seasons.

For example, the Cooper family hosts a seasonal gathering each fall to celebrate the harvest. They invite neighbors to help pick apples, gather vegetables from the garden, and then come together for a shared meal. The simplicity of the event allows for meaningful conversations and a sense of accomplishment, while also celebrating the fruits of their labor. By incorporating the seasons into your gatherings, you create a natural structure for your community events and can maintain a sense of rhythm throughout the year.

3. Creating a Welcoming Space

A minimalist home or gathering space is not defined by ornate decorations or excessive clutter but by warmth and openness. When hosting a gathering, the goal is to create an atmosphere that fosters connection, comfort, and relaxation. Simple touches—such as cozy seating, a well-set table, and a warm fire—can make guests feel welcome and comfortable without the need for excessive material possessions.

When hosting gatherings, you don't need to have a fully furnished or decorated home. Instead, focus on creating an environment that reflects your values. A few candles, a simple homemade meal, and some music in the background can go a long way in setting the right tone for your gathering.

The Power of Shared Tools, Skills, and Meals

One of the most valuable aspects of a minimalist community is the sharing of tools, skills, and meals. These simple acts of generosity and cooperation not only strengthen relationships but also reduce the overall burden on each individual homesteader. Instead of accumulating excessive amounts of tools or resources, minimalist homesteaders can rely on a strong network of like-minded people who share what they have and offer their skills freely.

1.   Sharing Tools

Tools are often a significant investment, and minimalist homesteaders recognize the power of sharing resources within their community. Instead of each family owning specialized tools that are rarely used, homesteaders can share or swap tools, making them available to everyone who needs them.

For example, the Carter family has a neighbor with a tractor, which they use for heavy-duty tasks like tilling the soil or moving large piles of wood. In exchange, the Carters share their pruning shears and chainsaw for the neighbor's personal use. This sharing system works because each person has something to offer, and there's no need to purchase or store expensive equipment that only gets used a few times a year.

Community tool-sharing is not just about practicality—it also fosters a deeper sense of connection and trust. Knowing that you can rely on your neighbors for help when needed creates a strong support system and eliminates the need for excess consumerism.

2.   Exchanging Skills

Minimalist homesteaders often have specialized skills that can be shared within their community. Whether it's animal care, gardening, building, or preserving food, the value of skills-based exchanges cannot be overstated. By offering your expertise and learning from others, you can enhance your own homesteading capabilities while helping your neighbors.

For instance, the Jackson family is skilled in woodworking, and they often help other homesteaders build structures, like raised garden beds, small chicken coops, or shelves. In exchange, they receive help with gardening, animal care, or harvesting crops. This exchange of skills is rooted in mutual respect and trust, creating a sense of shared purpose and collaboration.

3.   Sharing Meals and Food

Sharing food is one of the oldest forms of community building. A meal brings people together, nourishes the body, and creates an opportunity for meaningful conversation. In minimalist homesteading, meals are often prepared from simple, local, and seasonal ingredients, making them an ideal way to share and celebrate abundance.

The Thompson family, for example, regularly hosts community dinners where they invite their neighbors to bring dishes made from locally sourced ingredients. Each meal is an opportunity to share the bounty of the season and exchange stories, recipes, and ideas. By focusing on simple, wholesome food, these gatherings reflect the minimalist values of nourishment, connection, and sustainability.

In addition to large gatherings, smaller, informal food exchanges—such as trading produce, homemade bread, or canned goods—help create a robust network of sharing. This exchange not only reduces food waste but also strengthens ties between families.

Conclusion

Community, when embraced through the minimalist homesteading lens, becomes a source of strength, support, and sustainability. By focusing on quality over quantity in relationships and embracing shared resources, skills, and meals, minimalist homesteaders can build meaningful, resilient networks. Simple gatherings foster a sense of belonging and connection without the need for material excess, while the sharing of tools, skills, and food enhances the overall well-being of the community. In the end, the minimalist approach to community encourages collaboration, generosity, and sustainability, providing a foundation for both personal fulfillment and collective success. By living simply, we can live deeply, in harmony with our neighbors, the earth, and ourselves.

Challenges, Fails, and Learning Curves

N o matter how thoughtfully you plan or how diligently you work, life on a homestead is full of challenges, unexpected setbacks, and things going wrong. It's an inevitable part of living a self-sufficient lifestyle, and embracing those failures and obstacles is one of the most valuable lessons minimalist homesteading has to offer. This chapter will explore how to navigate the inevitable challenges that arise on the homestead, how minimalism can help you let go of perfectionism, and how embracing simplicity can reduce overwhelm when life throws its curveballs.

When Things Break or Go Wrong (And They Will)

One of the most important things to remember as a minimalist homesteader is that things will break, go wrong, and not turn out as planned—and that's okay. Whether it's a broken tool, an unseasonable frost, or a failed gardening experiment, challenges are a natural part of the homesteading experience. Instead of allowing these setbacks to create stress or disappointment, embracing them as opportunities for growth is an essential part of minimalist living.

1.   The Reality of Homesteading: It's Not Always Easy

The romanticized vision of homesteading often overlooks the fact that things don't always go as planned. You might plant a garden, only for a late frost to kill your seedlings. Your chicken coop may get damaged in a storm, or your well may run dry. When you raise livestock, there are inevitable challenges like illness or injury. These things can feel overwhelming, especially when they are out of your control.

Take, for example, the story of Tim and Laura. They had worked for months to establish a thriving vegetable garden, only to have a hailstorm devastate the crops just before harvest. Rather than giving in to frustration, they took it as an opportunity to learn. They researched how to build better crop protection systems and adapted their planting schedule to suit their climate. Instead of seeing the storm as a failure, they saw it as part of the learning curve—a necessary step toward becoming more resilient homesteaders.

Homesteading challenges don't always have a clear solution, and that's part of the process. Life on a homestead is full of trial and error, and not everything will work as you hope. What matters most is how you respond to these challenges. By approaching them with an open mind and a willingness to learn, you'll gain valuable experience that will serve you well in the future.

2.   Learning from Failures

Failures, while frustrating, often provide the most valuable lessons. It's easy to learn from success—things that go right are a source of pride—but failure has a way of teaching us deep lessons about patience, perseverance, and problem-solving.

For instance, when Chris and Emily first started keeping bees, they lost several hives in their first year. Rather than giving up, they dug into beekeeping books, attended workshops, and connected with local beekeepers to understand what went wrong. They learned about hive maintenance, the importance of monitoring for pests, and the optimal conditions for healthy bees. By embracing their failures and learning from them, they ultimately became successful beekeepers, and they now teach others how to care for bees.

Failures are an essential part of growth, and they help develop resilience—the ability to bounce back from setbacks, learn from mistakes, and keep moving forward.

Letting Go of Perfectionism

In a world where success is often measured by flawless execution, minimalist homesteading offers an opportunity to let go of perfectionism. Perfectionism can be a burden, driving people to overwork themselves, set unrealistic expectations, and experience constant stress. Minimalism, on the other hand, encourages us to focus on what truly matters, let go of the need for perfection, and embrace the beauty of imperfection.

1.   The Pressure of Perfect Results

The desire for perfection is a common challenge that many homesteaders face. You may want your garden to be lush and productive, your animals to be healthy, and your house to be immaculate. But this level of perfection is often unrealistic and can lead to burnout, frustration, and even guilt when things don't go as planned.

Take, for example, a typical gardening scenario. You might start the season with big dreams of perfectly straight rows of vibrant vegetables. However, the reality may be that the soil isn't ideal, some plants don't grow as expected, or pests invade your garden. These imperfections can be frustrating, but they are also part of the process. Instead of fixating on perfection, minimalist homesteading teaches us to embrace the process of growth, accept setbacks, and work with the rhythms of nature.

2.   Embracing Imperfection

One of the core principles of minimalist living is to focus on what's essential and let go of the unnecessary. This applies to your expectations as well. Homesteading doesn't require perfection—what it requires is effort, patience, and a willingness to adapt. When you let go of the idea that everything must be flawless, you free yourself to enjoy the journey and celebrate the small successes along the way.

For example, the Turner family used to struggle with the idea of having the "perfect" homestead. They felt pressure to grow all their own food, have an immaculate garden, and never let anything go to waste. But over time, they realized that these ideals were draining their energy and robbing them of the joy of homesteading. They shifted their focus to enjoying the process of planting, harvesting, and caring for animals, rather than trying to achieve perfection. They also allowed themselves to embrace imperfection in their home—dusty shelves, unfinished projects, and all—recognizing that their value came not from a perfect home but from a well-lived life.

3.   Simplifying Expectations

Minimalism encourages us to ask, "What is enough?" When it comes to homesteading, this means defining your goals based on what truly aligns with your values, not what others expect of you or what the "perfect" homesteading lifestyle looks like in the media. For example, if growing all your own food is a priority, you may find that having a small, well-tended garden that meets your family's needs is "enough." There's no need to compete with other homesteaders or try to achieve an idealized version of the homesteading dream.

Instead of getting bogged down by perfection, minimalist homesteading teaches us to enjoy the process, accept imperfections, and celebrate the progress we've made. Perfection is an illusion—it's the effort, the learning, and the connection with nature that truly matter.

How Minimalism Helps Reduce Overwhelm When Life Gets Real

Homesteading, by its nature, involves a lot of hard work and responsibilities. There are daily chores, seasonal tasks, and unexpected obstacles. Without a minimalist approach, the sheer volume of tasks can become overwhelming. However, by simplifying your life and focusing on what truly matters, you can reduce stress and approach challenges with a clearer mind and greater sense of calm.

1.   Prioritizing Essentials

Minimalism helps you focus on what's essential, which is especially helpful when life gets chaotic. By paring down unnecessary tasks, projects, and expectations, you can create more time and space for what truly matters. For instance, you might realize that while having a perfectly organized pantry is nice, what really matters is having enough food to feed your family. The minimalist approach helps you prioritize tasks that contribute directly to your well-being, rather than getting lost in perfectionist goals that only add to your stress.

For example, the Johnsons decided to scale back on the number of crops they grew in their garden. Initially, they wanted to grow everything—vegetables, herbs, fruit trees—but after one exhausting growing season, they realized that they weren't eating everything they grew, and a lot of it went to waste. Instead, they narrowed their focus to just a few crops that they actually enjoyed eating and could use regularly. This shift reduced the burden of gardening and allowed them to focus on the essentials.

2.   Simplifying Systems

Minimalism is about reducing complexity. This can be applied to the way you manage tasks and systems on your homestead. Instead of feeling overwhelmed by a long to-do list or complicated routines, simplify your processes to make things more manageable. Simplified systems reduce the mental load and help you stay focused on what matters most.

For example, the Morales family used to have a complex system for rotating crops and organizing garden plots. However, they found that it took too much time and energy to plan and manage. They simplified by creating a small, easy-to-maintain garden and relying more on companion planting and natural methods of pest control, reducing the need for complicated systems, and making the work more sustainable.

3.   Embracing Flexibility

Minimalism teaches us to be flexible and adaptable. On the homestead, things don't always go according to plan—whether it's a failed harvest, unexpected weather, or a sudden illness. Minimalism encourages you to let go of rigid expectations and be open to changing plans when necessary.

When life gets real, the minimalist mindset helps you adjust without becoming overwhelmed. Rather than stressing about everything going wrong, you can focus on finding simple solutions and moving forward. This mindset fosters resilience, helping you bounce back from challenges and make decisions that support your long-term well-being.

For instance, the Parkers found that their garden was damaged by a heavy rainstorm, washing away their freshly planted seeds. Instead of lamenting the loss, they simply replanted a few of their hardiest crops and made a plan for next season. They recognized that farming isn't about perfection but about doing the best you can with the resources you have.

Conclusion

Challenges, setbacks, and failures are an inevitable part of the minimalist homesteading journey. The key is to approach them with an open mind, a willingness to learn, and a commitment to resilience. By letting go of perfectionism, embracing imperfection, and simplifying your systems and expectations, you can reduce overwhelm and maintain your focus on what truly matters. Life on the homestead is full of learning curves, but it is through these experiences that we grow, develop resilience, and deepen our connection with the land and our communities. In minimalist homesteading, it is not the flawless execution of every plan that leads to success—it is the ability to adapt, learn, and continue moving forward, no matter what challenges arise.

CHAPTER 15:
Rooted and Real – Living a Legacy of Simplicity

Minimalist homesteading is not just a lifestyle choice—it's a philosophy that embodies living with intention, simplicity, and sustainability. It is about creating a life that is deeply connected to nature, community, and purpose. For many homesteaders, this way of life becomes a legacy—something they pass down through generations. But sustaining this lifestyle long-term, embracing its deeper value beyond material wealth, and defining your own version of success are not always easy. It requires constant reevaluation, adaptation, and a deep commitment to the principles of simplicity. This chapter explores how to sustain minimalist homesteading for the long haul, the unique gifts it offers that money can't buy, and how to live a life rooted in values that last.

Sustaining This Lifestyle Long-Term

Minimalist homesteading, at its core, is about sustainable living—not just for the environment but for the individual and the family. Creating a self-sufficient homestead requires careful planning, ongoing effort, and the flexibility to adapt to the inevitable changes and challenges that come with rural living. However, with the right mindset and systems in place, minimalist homesteading can be a fulfilling and sustainable way of life for the long term.

1.   Building Resilience

One of the key components of sustaining a minimalist homestead for the long-term is resilience. Whether it's dealing with financial challenges, extreme weather, or personal setbacks, resilience allows you to weather the storms of life with patience, grace, and adaptability. Building resilience on the homestead often comes from experience, learning from mistakes, and finding solutions that support long-term stability.

For example, consider the Thompsons, who have been homesteading for 15 years. In their early years, they faced many challenges, from crop failures to equipment breakdowns. However, over time, they learned how to anticipate problems and build systems that would help them cope with difficulties. They implemented water conservation techniques, learned how to repair tools themselves, and developed relationships with neighbors for mutual support. By building this resilience, they've been able to sustain their homestead year after year without burning out.

2.   Adapting to Changing Needs

Minimalist homesteading is not a static lifestyle; it evolves over time. Whether it's due to aging, health concerns, or changes in family dynamics, the needs of a homestead will shift, and flexibility is crucial. For example, as children grow up and leave home, the homesteading routine might change to accommodate new priorities. Similarly, as climate or economic conditions change, homesteaders may need to adjust their methods of farming, gardening, or managing resources.

Sustaining a minimalist homestead long-term requires an openness to change and a willingness to adapt. This adaptability can be seen in the example of the Green family, who, after their children grew up, shifted from growing food for a large family to focusing on smaller, more specialized crops, which allowed them to better manage their time and energy. By continually reassessing their goals and needs, the Greens have been able to keep their homestead sustainable, simple, and manageable.

3.   **Prioritizing Self-Care and Balance**

Sustaining a minimalist homestead for the long term also requires maintaining personal well-being. While homesteading is often a fulfilling and satisfying lifestyle, it can be physically demanding and mentally taxing. Setting boundaries, taking time for rest, and finding ways to recharge are essential for avoiding burnout. Minimalist homesteading teaches us that rest is not a luxury—it's a necessity for long-term sustainability.

Self-care could be as simple as taking a few moments each day for mindfulness, ensuring that you have time for hobbies or creative pursuits outside of homesteading, or planning regular breaks from work to spend time with loved ones. By maintaining balance and making self-care a priority, you'll be better equipped to handle the demands of homesteading and continue living simply for years to come.

What Minimalist Homesteading Gives You That Money Can't

One of the most profound realizations for minimalist homesteaders is that the lifestyle offers a sense of fulfillment and wealth that money simply can't provide. While financial independence and material comfort are often valued in modern society, they can't replace the deep sense of purpose, satisfaction, and connection that minimalist homesteading provides.

1.   Connection to Nature

One of the most priceless aspects of minimalist homesteading is the deep connection to nature. When you live off the land, care for animals, and grow your own food, you develop an intimate relationship with the cycles of nature. This connection fosters a sense of awe, wonder, and gratitude for the earth's resources. It is a form of wealth that no amount of money can replace.

For example, Emma, who lives on a small homestead in the Appalachian mountains, wakes up every morning to the sounds of birds and the sight of mist rising from the valley. She spends her days working in the garden, collecting eggs, and harvesting herbs. The deep satisfaction she feels in these simple moments is a form of wealth that cannot be measured by dollars.

2.   Self-Sufficiency and Independence

Minimalist homesteading also gives you the freedom to live independently. Instead of relying on external systems, such as grocery stores, power grids, or even healthcare, you have the ability to provide for yourself and your family. This sense of self-reliance gives you a freedom that money cannot buy.

For instance, the Williams family made the decision to go off-grid and build their own solar system. While it was a significant investment of time and effort upfront, they now enjoy the peace of mind that comes with knowing they are not dependent on external sources of power. The satisfaction of being able to meet their own needs gives them a sense of security and autonomy that cannot be found in a paycheck.

3.   Meaningful Work

Minimalist homesteading offers the opportunity to engage in meaningful work—work that is intrinsically rewarding and aligned with your values. Whether it's raising animals, tending to your garden, building structures, or crafting goods, the work you do on the homestead is directly tied to your well-being and the well-being of those around you. This work provides a sense of purpose that cannot be purchased.

Consider the story of Peter and Laura, who moved to their homestead to escape the corporate grind. Initially, they struggled financially, but over time, they found that the work they were doing—growing food, caring for animals, and building their own home—was fulfilling in a way that no paycheck could match. The meaningful nature of their work gives them a sense of satisfaction and accomplishment that money simply can't provide.

Writing Your Own Definition of Success

Minimalist homesteading challenges traditional definitions of success. In a world that often defines success in terms of wealth, status, and career achievement, minimalist homesteaders have the opportunity to rewrite what success means to them. Success, in the context of minimalist homesteading, is not about accumulating wealth or climbing the corporate ladder—it's about creating a life that aligns with your values, fosters personal fulfillment, and contributes to the well-being of your family, community, and the environment.

1.   Redefining Success

For minimalist homesteaders, success is about finding balance, living intentionally, and being content with enough. It's about having the freedom to choose how you spend your time, knowing that you are living in alignment with your core values. It's about creating a sustainable life that is rich in experiences, relationships, and connection to the land.

For example, Maria and Sam chose to leave their corporate jobs and move to a small homestead in the countryside. While they are not wealthy by conventional standards, they consider themselves successful because they have created a life where they can focus on what truly matters—family, nature, and meaningful work. To them, success is about living authentically, rather than chasing external accolades.

2.   Living by Your Values

Minimalist homesteading allows you to live by your values rather than societal expectations. By simplifying your life and focusing on what's truly important, you can create a legacy that reflects the things that matter most to you, whether that's sustainability, self-sufficiency, creativity, or community. Defining success on your own terms helps you stay grounded and focused, regardless of the pressures or comparisons that may arise.

Final Thoughts: Choosing Simplicity Again and Again

Minimalist homesteading is a journey—a journey that requires continual reassessment, adaptability, and a deep commitment to simplicity. It's a path that will lead to both challenges and joys, but ultimately, it will be one of personal fulfillment, connection, and purpose. Embracing simplicity isn't a one-time decision; it's a continual practice—choosing simplicity again and again, even when life gets complicated.

As you move forward on your homesteading journey, remember that minimalism is about making intentional choices that align with your values. It's about living in harmony with nature, focusing on what truly matters, and letting go of the excess that distracts from your goals. By choosing simplicity, you create the space to build a life that is rooted in authenticity, purpose, and deep satisfaction—a life that, over time, will become your legacy of simplicity.

Ultimately, minimalist homesteading offers a way of life that is more fulfilling than anything money can buy. It's a life of resilience, connection, and meaning. By continuing to choose simplicity, you ensure that this way of life becomes not just a phase, but a lasting legacy—one that will endure for generations to come.

Conclusion

In the end, minimalist homesteading is about much more than just living with less—it's about living with intention. It's about creating a life that is rooted in meaningful work, simplicity, and connection. As you strive to sustain this lifestyle long-term, you realize that the value it brings isn't something that can be measured by wealth or material possessions. What minimalist homesteading offers you that money cannot buy is the gift of self-sufficiency, the freedom to define your own success, and a deep, fulfilling connection with nature and your community.

Conclusion
Embracing a Legacy of Simplicity through Minimalist Homesteading

The journey through minimalist homesteading is not just about a lifestyle change; it's a philosophy that influences how we live, work, and interact with the world around us. From the first chapter, where we explored the call to simplicity, to the final one, where we reflect on the legacy we build, the essence of minimalist homesteading is about simplifying life while deepening our connection to nature, community, and purpose. This journey leads us to a richer, more meaningful existence—one that thrives on less but flourishes in value.

In the beginning, we examined why people are choosing minimalist homesteading, recognizing that it is a reaction against the overwhelming complexity and consumerism of modern life. Instead of chasing after fleeting desires, minimalist homesteaders turn to the land and simple, sustainable practices that allow them to take control of their lives. They learn to live with intention and purpose, cutting through the clutter to focus on what truly matters: relationships, nature, and well-being.

As we moved through the chapters, we learned about designing a simple homestead, understanding that creating a life of simplicity doesn't mean abandoning comfort or beauty but involves a mindful approach to choices. Whether it's choosing land that aligns with your values or creating spaces that serve, not stress, the minimalist homesteader focuses on functionality and sustainability, emphasizing the beauty of modesty and the power of intentional living.

In the chapters above, we dove deeper into the practical aspects of homesteading, from managing energy systems like solar and wind, to learning how to grow food with intention, and how to simplify the kitchen for nourishment. These chapters revealed how minimalism thrives on efficiency, resourcefulness, and sustainability. By integrating thoughtful choices and efficiency into each aspect of life—whether it's how we cook, grow, or use energy—we simplify daily tasks and embrace a life that is rooted in respect for the environment and its resources.

We then took a closer look at the tools, skills, and mindset needed for a fulfilling homesteading life. Learning to make and mend things, from repairing tools to sewing, empowers us to live with greater independence. This approach not only reduces our reliance on external systems but also strengthens the sense of accomplishment that comes from doing things ourselves. The minimalist homesteader doesn't fear challenges or failures; instead, they embrace them as opportunities for growth, understanding that mistakes are part of the learning process.

One of the core tenets of minimalist homesteading is the value of community—how we find and build connections, share tools and skills, and exchange meals with others. Community provides a strong foundation upon which we can rely during times of difficulty, and minimalist homesteading emphasizes creating connections that are meaningful, based on shared values of sustainability and simplicity. These connections nurture relationships, foster a sense of belonging, and remind us that we don't have to navigate the challenges of life alone.

Throughout these chapters, the recurring theme has been the understanding that minimalism offers more than just fewer things—it provides a deeper connection to what matters most. As we discussed the power of challenges, failures, and learning curves, we acknowledged that life on a homestead isn't always easy or perfect. But through resilience, adaptability, and the courage to face setbacks, we can build lives that are rich in meaning and free from the pressure of perfectionism. Minimalism teaches us to let go of the need for flawless outcomes, and instead, embrace the simplicity and beauty of life as it unfolds.

Finally, the minimalist homesteading journey leads us to the legacy of simplicity that we leave behind. Minimalist homesteading offers something money can't buy: a life of self-sufficiency, deeper relationships, and an enduring connection to nature. As we learn to live more simply, we redefine success— not as the accumulation of wealth, but as a life aligned with our core values, one that fosters contentment and well-being for ourselves and future generations.

This lifestyle, rooted in intentional living and simplicity, is a continual process of choosing again and again the values that align with who we are and who we want to be. It is a call to slow down, focus on what matters, and live with gratitude for the abundance that nature and simplicity can provide.

Nicci Brochard & Dr.Ben Chuba

Minimalist homesteading is not just about reducing clutter or getting back to the basics; it's about living more fully, more authentically, and more in tune with the rhythms of nature. It is about creating a life that reflects the beauty of simplicity, and in doing so, building a lasting legacy that thrives on what is essential, meaningful, and deeply rooted. This is the true gift of minimalist homesteading: a life that is rooted and real, a life worth living and sharing.

Epilogue

The journey has just begun. As the final embers of another handmade day fade into dusk, *Simple Roots* leaves you not with a checklist, but a choice. In embracing the quiet rhythm of soil, seasons, and self-sufficiency, you've discovered that less truly *is* more.

This journey is about your presence. The modern homestead, whether a sprawling acreage or a balcony of herbs, begins with intention. Let your hands stay calloused, your pantry full, and your heart lighter for having walked a simpler path. Return often to these roots, not out of nostalgia, but necessity. In simplicity, we grow. In stillness, we remember. And in choosing less, we find, at last, enough.

Acknowledgement

Writing *Simple Roots: The Minimalist's Guide to Modern Homesteading*, has been a deeply personal and transformative journey for myself (Nicci) and for Ben. This book would not have been possible without the support, patience, sacrifice, and inspiration of many people.

We hope you enjoyed the book. Kindly leave us a review.

Nicci & Ben

www.ingramcontent.com/pod-product-compliance
Lightning Source LLC
Chambersburg PA
CBHW061807120626
46550CB00005B/2169